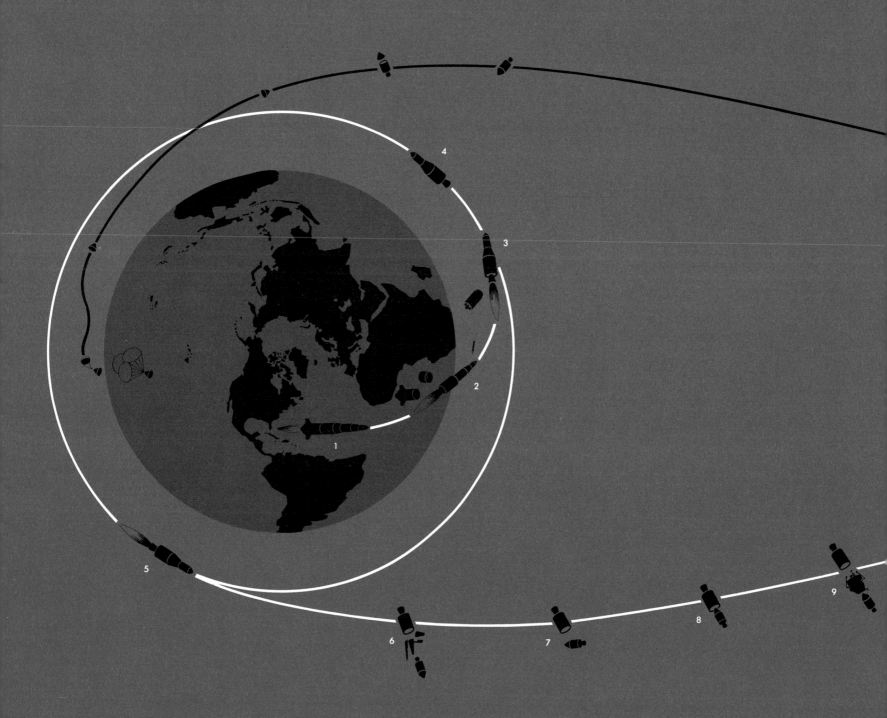

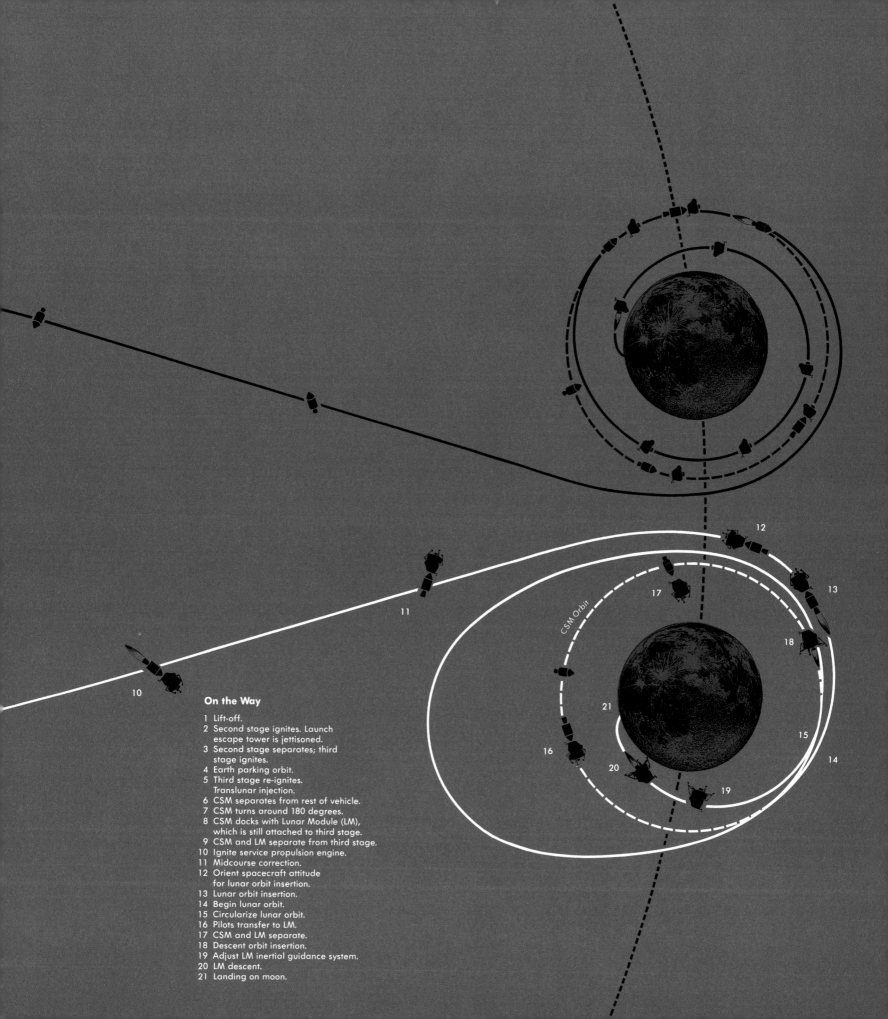

On the Way

1 Lift-off.
2 Second stage ignites. Launch escape tower is jettisoned.
3 Second stage separates; third stage ignites.
4 Earth parking orbit.
5 Third stage re-ignites. Translunar injection.
6 CSM separates from rest of vehicle.
7 CSM turns around 180 degrees.
8 CSM docks with Lunar Module (LM), which is still attached to third stage.
9 CSM and LM separate from third stage.
10 Ignite service propulsion engine.
11 Midcourse correction.
12 Orient spacecraft attitude for lunar orbit insertion.
13 Lunar orbit insertion.
14 Begin lunar orbit.
15 Circularize lunar orbit.
16 Pilots transfer to LM.
17 CSM and LM separate.
18 Descent orbit insertion.
19 Adjust LM inertial guidance system.
20 LM descent.
21 Landing on moon.

CSM Orbit

45

5/86

10:56:20PM EDT
7/20/69

*The historic conquest of the moon as reported to the American people
by CBS News over the CBS Television Network.*

Contents

Foreword

At 10:56:20 p.m. EDT, July 20, 1969, man first stepped on the moon. It took place 238,000 miles out in space, yet it was shared by hundreds of millions of people on earth. The step on the moon was an awesome achievement; so was its reporting on television because it emphasized television's extraordinary ability to unify a disparate world through communicating with so many people, in so many places, and thus providing them with a common—and an extraordinarily satisfying—experience.

Like all experiences recorded by television, this one was ephemeral. It flashed on the screen and was gone. This book, which attempts to recapture some of the reality of this unique experience, also permits all of us, in some small measure, to recall that experience in more permanent form. In this sometimes discouraging and frustrating world, it is worth preserving a moment of pride in the ability of man to do what he sets his mind to do. It not only serves as a mark of what man has done, but what, with similar determination and cooperative effort, he can do in the future to achieve other goals, closer to home. And so, this book not only seeks to crystallize the images that CBS News brought to the screen, but recounts the story of the intricate logistics and inventive techniques of the men on camera and behind the camera who made this reality possible.

For my colleagues at CBS News, and for myself, covering "Man on the Moon: The Epic Journey of Apollo 11" ranks as the single most satisfying effort in our collective experience as journalists. All too often we are forced to report man's shortcomings. In this instance, from the moment of blast-off to the moment of splashdown we were continually conscious of being involved in one of the great triumphs of the human spirit.

This consciousness involved our total energies and efforts. It enabled us to meet some of the most formidable challenges ever faced by electronic journalism. Because it was one of man's greatest achievements, it was one of television's great achievements. And just as Apollo 11 could not have been achieved without the extraordinary genius and dedication of hundreds of thousands of Americans who stood behind the three astronauts, so the CBS News broadcast could not have been achieved without the magnificent dedication and excellence of the hundreds of people who make up CBS News.

RICHARD S. SALANT
President, CBS News

The Launching
July 16

Wednesday, July 16, 1969

Mission Control: We are still go with Apollo 11…30 seconds and counting. Astronauts reported "Feels good." T-25 seconds. 20 seconds and counting. T-15 seconds, guidance is internal, 12, 11, 10, 9, ignition sequence starts, 6, 5, 4, 3, 2, 1, zero, all engines running, lift-off. We have a lift-off, 32 minutes past the hour. Lift-off on Apollo 11. Tower cleared.

It is 9:32 a.m. EDT on Wednesday, July 16, 1969. Millions of Americans and countless millions around the world have waited for this moment. Astronauts Neil Armstrong, Edwin "Buzz" Aldrin and Michael Collins are riding a huge Saturn V rocket, lifted by 7,500,000 pounds of thrust, on their way to a date with history.

At lift-off, the CBS News studio at Cape Kennedy, which only five minutes before had been a scene of great activity, was virtually empty. Only Correspondent Walter Cronkite, former astronaut Wally Schirra, producer Joan Richman, stage manager David Fox, the cameramen and two technicians were in the studio as the countdown entered its final minutes. Everyone who could had gone outdoors to watch the launch.

At the precise second of ignition all attention turned to the launch pad, and everyone moved to the window for a better view of the drama unfolding some three miles away. The window affords a panoramic view of the press site and the pad. Miss Richman was struck by the size of the crowd. Never in her memory had so many people been at the Cape for a launch. The edge of the jetty in front of the press site was ringed with people, and the overflow of cars from the normally adequate parking lot had spread onto adjoining roads and into fields surrounding the viewing area.

As great a vantage point as the window offers for lift-off, once the rocket clears the tower it disappears from view. Miss Richman remembers lying flat on her back on the floor to watch the huge Saturn V as it arched out over the Atlantic Ocean. She was "sort of entwined with Walter's chair legs," in order to see it pass by the window frame. Schirra had dropped to his hands and knees on the floor, with his face pressed against the glass at what Fox described as a 180-degree angle. He gave Miss Richman and Cronkite the thumbs up sign and said, "There she goes. It's beautiful!"

Cronkite, watching on a monitor in the studio after the rocket had left his view, was speechless during the early seconds of lift-off. Then as he saw the rocket on its way, he jubilantly cried, "Oh boy, oh boy, it looks good, Wally. Building shaking. We're getting the buffeting we've become used to. What a moment! Man on the way to the moon! Beautiful!"

Man was on his way to the moon, fulfilling an age-old dream that he would someday leave his home planet to stand in lonely splendor on a celestial body.

On the eve of America's embarkation on this great adventure CBS News National Correspondent Eric Sevareid spoke eloquently about the space program and the American spirit.

Sevareid: *The great debate about America in space is an exercise in freedom, the freedom of choice. How shall a people use its excess energies and resources?*

Here at the launch site for our most spectacular adventure, the doubts, and they are minor doubts, concern the success of this flight and the safety of the astronauts. In Washington and elsewhere, the doubts concern future flights, their number, their cost and their benefits, as if the success of Apollo 11 were already assured.

We are a people who hate failure. It's un-American. It is a fair guess that the failure of Apollo 11 would not curtail future space programs but re-energize them. Success may well curtail them because for a long time to come future flights will seem anti-climactic, chiefly of laboratory, not popular, interest, and the pressure to divert these great sums of money to inner space, terra firma and inner man will steadily grow.

The core of the argument is philosophical, almost religious, and concerns the nature of man and his society. There are men here pas-

sionately convinced that we cannot plan our opportunities but must take them at the tide when they come, that without new frontiers always beckoning, an energetic people rusts and corrodes, that space is not only the cutting edge for science today but a moral substitute for war that could give the quarreling human race some sense of common identity, of brotherhood.

And there are men convinced with equal passion that this adventure, however majestic its drama, is only one more act of escape, that it is man once again running away from himself and his real needs, that we are approaching the bright side of the moon with the dark side of ourselves, with chauvinism, competitive aggression, and that the chief uses of space will be military in the end.

We do know that history has never proceeded by a rational plan. Not even science knows what it is doing much beyond the immediate experiment. We know that the human brain will soon know more about the composition of the moon than it knows about the human brain. We will know why the moon is what and where it is before we know why human beings do what they do. We are still closer to our animal origins than to any millenium in peace.

It is possible that the divine spark in man will consume him in flames, that the big brain will prove our ultimate flaw, like the dinosaur's big body, that the metal plaque Armstrong and Aldrin expect to place on the moon will become man's epitaph. But the future has never revealed itself. It is made, step by step, moment by moment. The next great moment is now at hand. The world waits and the astronauts sleep.

Wednesday, July 16, was an unusual day for most of the CBS News "regulars" at the Cape. A few, like David Fox, and of course Cronkite, who was covering his twenty-first manned flight, had been making these trips to Florida since 1961. Whereas 40 people were the usual complement for an Apollo shot, the work force had been increased three- or four-fold to cover Apollo 11. CBS President Frank Stanton was on hand with CBS News President Richard S. Salant and Peter Goldmark, President of CBS Laboratories, the man who developed the field sequential system used in the color television camera aboard Apollo 11.

In addition to Cronkite and Sevareid, other CBS News correspondents reporting on various aspects of the story were David Schoumacher, who regularly covers the space programs, Heywood Hale Broun, Ike Pappas and Ed Rabel. Schirra, as a special analyst, was joined by Arthur C. Clarke, author of *2001: A Space Odyssey* and one of the world's leading space visionaries, and Dr. Newell Trask, a lunar mapping specialist from the United States Geological Survey. The CND Wire, an independent, in-house news-gathering organization that is put into operation during the coverage of major news events, was in full swing, filing stories on the progress of the mission, and information of general news interest about the people involved and the events to come in the next eight days.

The day had begun at 4:00 a.m., as everyone gathered at the press site to prepare for broadcasting at six o'clock. These were restless moments. The service structure that shrouds the rocket had been pulled away, and most people were seeing the huge Saturn for the first time. Great floodlights played on the launch pad and rocket.

At 6:00 a.m., the broadcast day began with the announcement from New York: "This is a CBS News Special Report, 'Man on the Moon: The Epic Journey of Apollo 11.'" Cronkite was in position at the anchor desk, beginning the first of 46 hours and seven minutes of coverage to be devoted to Apollo 11 over the next eight days. With Schirra at his side, he set the scene for the lift-off, which was now three hours and 32 minutes away.

At 6:25 a.m., all attention turned to television monitors showing the exterior of the Manned Spacecraft Operations Building. The crew transfer van that would carry the astronauts to the launch pad sat in the driveway, and at 6:27 the door opened and a smiling Neil Armstrong led Michael Collins and Buzz Aldrin down a ramp and into the van for their 10-mile trip to Launch Complex 39A.

The van drove up the road alongside the press site, and most people kept their eyes on the monitors checking its progress so they could go outside and watch as it passed by. On this morning, the astronauts who were to travel as fast as 25,000 miles an hour during their trip to the moon were delayed by a good old-fashioned traffic jam. The usual 12 minutes from when they left the building to the time when the first astronaut entered the Command Module stretched to 27 minutes.

Eric Sevareid, who was seeing his first manned shot, described those early hours and his reactions in a talk with Cronkite shortly before lift-off.

Cronkite: *What's the view out there from the outside, Eric?*

Sevareid: *Walter, the heat is beginning to rise. Some slight haze gathering in the atmosphere. You don't quite see the rocket out there as distinctly as we saw it shortly after dawn. The heat is like a silk cloth you put across your face. Before dawn the rocket out there was all lit up, pure white. It looked like the North Pole would look if it was really a pole, with lights of the aurora borealis going out in every direction in the black skies around.*

The whole country around here is a kind of vast launching pad. This is big sky country. It's very flat. It's right-angle country. And the crowd is always knots of people. There's not a carnival atmosphere here really. You've got the snack shops and all the rest, all the trailers, but there is a quiet atmosphere and when the vans carrying the astronauts themselves went by on this roadway just now, there was a kind of a hush among people. Those things move very slowly as though they're carrying nitroglycerine or something. You get a feeling that people think of these men as not just superior men but different creatures. They are like people who have gone into the other world and have returned, and you sense they bear secrets that we will never entirely know, that they will never entirely be able to explain.

There are quite a few children around, not too many in this area, but all along the beaches and roadways. I think they look at this a bit differently too, Walter. It is probably not as awesome to them as it is to people of your generation and mine. They've sort of grown up with this. It's all been anticipated.

You remember when you saw your first airplane? I did out on the prairies in Dakota and it was a tremendous thrill. I wonder if it's the

13

same for these kids. I don't think the past means anything to them. This is all very natural to them.

Cronkite: *I've noticed in the reporting that those under 16, who've really grown up with space since its first memorable moments—when they were four or so and the space thing was just coming into being—understand it and want more detail in our reporting. They want to know about escape velocity and they want to know about the lunar trajectory velocity, and those over 30 or so say, "Don't tell me all that, I just don't understand. Tell me when we get there."*

Sevareid: *Furthermore, this is not a romantic era, not a poetic era. The beauty the young find is in the things themselves. All the imagery and the words will come later, but we really don't have a language to describe this thing.*

As we sit here today, what are the words you use? I think the language is being altered, many new words and phrases and concepts are being added, and, I think, some language is being eliminated. How do you say "high as the sky" anymore, or "the sky is the limit"—what does that mean?

Cronkite: *Maybe it's that we have been so busy, so many things crowding in on us, we haven't had time for language.*

Sevareid: *There's always a great cultural lag on these things. It takes a long time for a new language to emerge.*

Many of the spectators had never witnessed the launch of a manned Apollo mission. For them, the moment of lift-off was an extraordinary one.

Treated pretty much like rookies who are about to play their first major league baseball game, they were regaled with stories of how they were about to witness one of the most awesome sights known to man. They were told that the man-made explosion is second only to that of the atomic bomb, that the roar at lift-off is deafening and the flames blinding, and that you can actually feel the sound waves slap against you as the Saturn V climbs the tower.

These descriptions aren't far from wrong. On a good, clear day one can see the "bird" some two and one-half minutes into its flight, two and one-half minutes that seem like twice that long to even the most casual observer. It seems that it takes a minute before the rocket starts to move, and an hour before it starts to climb the tower, and then streaks off into the sky spitting a white flame back at those on the ground.

You wait for the word that everything is going as planned. Then it comes from the astronaut serving as capsule communicator (CAPCOM) in Houston.

Capcom: **This is Houston, you are go for staging.**

Apollo 11: **Inboard cutoff.** [The inboard engines on the first stage have been shut down.]

Capcom: **Inboard cutoff.**

Apollo 11: **Staging and ignition.** [Astronaut Neil Armstrong tells the ground that the first stage of the rocket has fallen away and that the engines of the second stage have ignited on schedule.]

Capcom: **11, Houston, thrust is go all engines, you are looking good.**

Apollo 11: **Roger. Hear you loud and clear, Houston.**

Everyone reacts differently during these moments. Common symptoms are a sudden cold feeling in the chest and tears in the eyes, even for those who have lived the experience before. Eric Sevareid admitted that his eyes filled with tears when he saw Apollo 11 leave the launch pad. Dr. Ralph Abernathy, who led a poor people's march to the area to protest the huge expenditures of money in space that he believed should be spent on eliminating poverty, told CBS News Correspondent Ed Rabel that for a few moments he forgot why he was there.

Abernathy: *There's a great deal of joy and pride. For that particular moment and second I really forgot that we have so many hungry people in the United States of America. But now I remember that we will have to go back to business as usual in trying to really launch a pro-*

15

gram that will move off on schedule and with the speed and rapidity as did this marvelous and magnificent rocket. I was one of the proudest Americans as I stood on this soil, on this spot. I think it's really holy ground and it will be even more holy once we feed the hungry and care for the sick and provide for those who do not have homes.

Robert Wussler, executive producer of the Apollo 11 coverage, was one of the most fascinated spectators. Working in the control room, under the studio, he watched the launch through a window much smaller in size than that in the studio. It was an unusual experience for Wussler who had been involved in the space program since its earliest days, long before Alan B. Shepard, Jr. went on the first suborbital flight. Wussler had never seen an Apollo launch before.

Wussler was at the Cape because the decision had been made several months before that this lift-off was of such importance that all of the "Launch Day" coverage should be controlled from Florida. On prior Apollo missions the press site coverage had been a "remote," with Wussler running the show from New York. Wussler stayed in the control room through lift-off, although at the last moment he went to the window with his camera to take a few pictures of Apollo 11 as it left the launch pad. "I wanted to be able to say," he remarked later, "that these were the pictures I took on the day man left for the moon."

Wussler had had little to worry about that morning. Everything was going well. There had been a couple of small problems, such as special phones failing to work and clocks stopping, but those were the usual ones encountered at the Cape. The one thing that was unusual was the failure of NASA's launch pad microphones. These are the mikes that capture the sounds of a launch. It was not until five minutes before lift-off that NASA announced that they were not working and that instead of the natural sound they would provide a recording of the Apollo 10 lift-off, synchronized with the ignition of Apollo 11. That offer brought a polite, "No, thank you," from CBS News President Richard Salant, and CBS viewers heard the sound of the launch through CBS microphones placed in front of the studio. The roar may have seemed a little late to some people, but they were hearing it just as everyone at the press site did.

The relative calm turned to bedlam within minutes of lift-off. CBS News had extended invitations to both Vice President Agnew and former President Johnson to come to the studio to talk to Cronkite. Both accepted, but both accepted for 10 o'clock. The former President and his party arrived first and were taken to a reception area behind the studio where he was met by Stanton and Salant. His car was still in the driveway when a second limousine, this one carrying the Vice President and astronauts Tom Stafford and Bill Anders, pulled up. The Vice President was taken directly to the studio. While his party, which numbered about 20 people, squeezed into the back of the studio, the Vice President talked with Cronkite about the launch of Apollo 11 and the future of America's space program.

Agnew: *I'm glad to be here, Walter.*

Cronkite: *So good to see you, sir. And this was quite a launch, wasn't it?*

Agnew: *Indeed it was. Each one of them is quite a launch, but I think the more you see the more exciting they get. First one I've seen from the outside. It's even more exciting out there.*

Cronkite: *Did anything about the launch surprise you?*

Agnew: *Just that it seemed to go easier than the other two. I think you get to learn a bit about the things that make you apprehensive—like the "lean-out" at start. It scared the dickens out of me the first time.*

Cronkite: *I know. And the slow climb, too, I think is frightening the first time you see it. Even though you know it's going to be that way, you just can't believe it's really moving.*

Agnew: *You get that sense of waiting for something to take off quickly, and it doesn't happen.*

Cronkite: *I know, but it was a beautiful sight.*

Agnew: *It was. It was indeed. And I'm just filled with a real feeling of great pride for these people, not just the three men that are in earth* 17 *orbit now, but the people behind the program. They're just so dedicated. I see a great future for this program.*

Cronkite: *You know, it's the nature of the American, and the people in the space program particularly, to constantly look beyond where we are. This is the nature of the man who wants to go to the moon. You were reported as saying—and everybody's looking for what you're saying as an indication of what this Administration's intentions will be toward space and beyond the moon—well, let me read the quote, and let's talk about it a moment. You said, "I think the United States should undertake a very ambitious new project in space. I think we shouldn't be ashamed to attempt something even though the scientific probability is in doubt. I think we should attempt interplanetary exploration in a manned sense." Do you still think that?*

Agnew: *I don't think we'd be out of line in saying we are going to put a man on Mars by the end of the century. And I think we should do it because, based on the rate of progress that we've shown, I think it's possible. And I think the people in the country, the average man, wants something to look forward to as an exciting objective. And in spite of all the objections about the spending of money, the space program will probably turn out to be one of our best investments, in time.*

By the time the Vice President left and former President Johnson prepared to go to the studio, most of the 3000 people at the press site were aware of their presence, and the area around the CBS News building was swarming with reporters, photographers and those attracted by the crowd. As soon as the Vice President left, Mr. Johnson was ushered to a seat next to Cronkite.

David Fox recalls a brief moment of "panic," when he was unable to put a microphone on the former President before the interview began. It was attached on camera, while Cronkite told Mr. Johnson, "I think I greeted you before you were attached up. There you are. I think you're wired for sound now, sir."

Cronkite: *You said at lunch, yesterday, Mr. President, that you had*
flown along on every one of these missions. But those you watched on

your television monitor at the White House, and this one you saw for the first time in person. Awesome sight, wasn't it?

Johnson: *It certainly was, Walter. It's a great thrill. I had the feeling of great concern for the outcome of this flight. We haven't reached the end. It's just the beginning, and it's been a long time in going as far as we have. The decision was actually taken 12 or 13 years ago that made possible that awesome sight this morning, when President Eisenhower put an extra hundred million dollars in a very tight budget, back in 1958.*

But you never get the feeling on one of these blast-offs, sitting in a room watching them on camera, that you get seeing them in person. And as they took off there this morning, I thought about how fortunate we'd been all these years to have had a minimum of accidents. And I know that all of our people are going to have great concern until this flight is finished.

Another reaction I had was that it just seemed like a half-a-million people who'd worked on this program through the years, each of them was there, just lifting their all, and trying to see that great power going to the skies.

Cronkite: *We say that if we can spend 24 billion dollars to get to the moon, we can do anything. How do we translate that into action?*

Johnson: *Back in '58, I urged President Eisenhower to say to the other nations of the world, "Let's all join in a united space program." We haven't been able up to now, to get other nations to agree. But President Nixon's administration very shortly will be engaged in negotiations and discussions with other leaders. It may be that more will come out of this than we know now.*

Cronkite: *Well, it's certainly something that we can rest our hopes in, even as we rest our hopes in those three men aboard Apollo 11.*

Johnson: *Well, as we walked away this morning, I thought of three things. I felt very deeply concerned for the men and their safety. I felt the great awe for what I'd just seen as they took off. And something you don't hear much about these days, I felt a great pride in this country and its ability to set aside partisanship, and differences, and quarreling among its scientists, and its industry leaders, its government* 19 *leaders. If there's been any of that, it's been held to a minimum. But if our industrial people, these great managers of industry, the laboring people, the government, the scientists, all with the help of the Congress can get together and do a job like this, there's just not anything we can't do. And there's so much that we have yet to deal with. The hunger in the world, the sickness in the world, the poverty in the world. We must apply some of the great talents that we've applied to space to all of these problems, and get them done, and get them done in the spirit of what's the greatest good for the greatest number.*

I think where we're better than anyone could suspect is the openness in this entire program. Everything we've done, we have done it with the tourists going through the space program, seeing everything with the media cooperating, everything in front of the camera. There's no secrecy.

Cronkite: *This has been incredible this week. You come out here to shoot a press story, a film story at the pad, and here are the buses rolling up with 50 people aboard from all parts of the United States and the world, getting out and looking around the pad. This was wide open as it can be.*

Johnson: *That's something our system has that no other system has equalled, and I think that's one of the reasons we have the strength we do.*

Cronkite: *No doubt about that, sir. Thank you, very much, Mr. President.*

An estimated half million people had converged on the area to view the launch in person. One of the most popular locations was along the beach in Cocoa Beach, some 15 miles south of the launch site. A few of the people had been there for two or three days, staking out a little area for themselves, and most had spent the night on the beach. Heywood Hale Broun was with them on the beach and described their reactions at the moment of lift-off to Cronkite.

20

Broun: *Walter, when the moment came which everybody had been waiting for, it seemed to stun them into a kind of frozen disbelief. They couldn't quite believe that man was finally on his way to worlds outside the one where he began. And as it rose higher and higher it began finally to move the eyes upward. At a tennis match you look back and forth. On a rocket launch you just keep going up and up, your eyes going up, your hopes going up, and finally the whole crowd like some vast many-eyed crab was staring out and up and up and all very silent. There was a small "Aah" when the rocket first went up, but after that it was just staring and reaching. It was the poetry of hope, if you will, unspoken but seen in the kind of concentrated gestures that people had as they reached up and up with the rocket.*

Enthusiasm on that morning was not reserved for spectators who were seeing their first launch. Arthur C. Clarke, who acted as a special analyst during CBS News' coverage of Apollo 10 in May, was back for Apollo 11. He had envisioned this day many years before, and in *2001: A Space Odyssey* had given his version of what life would be like in the 21st Century. Clarke had viewed the launch from the area in front of the studio, and later had joined Cronkite to talk about the lift-off and man's future in space.

Cronkite: *Arthur Clarke, I've got you back here in the copilot's seat with me, I'm pleased to say. What are your observations on the flight of Apollo 11 up to now?*

Clarke: *Well, it is one of the most thrilling things I've ever seen. I still have in my mind the image of that beautiful ship going up; and it's curious, although it's three miles away it seems enormous, it seems as though it's only a few hundred yards away. Where it rises the rest of the landscape vanishes. In my memory I can only see the rocket filling the sky, although of course I know it was only a fairly small visual image.*

Cronkite: *Arthur, we had some interesting comments from Vice President Agnew a little earlier about man going on to other missions in*

21

space. What do you see for the remainder of this century? Give me a sort of a time line as to how you would see the development in space from here.

Clarke: *Well, in the next ten years you're going to have two separate developments: the establishment of manned orbiting stations, space labs and perhaps the first primitive space factories; simultaneously we'll see the development of the first semi-permanent and permanent bases on the moon, comparable to those in the Antarctic today. Both these things are going to happen within the next ten years, in fact probably within the next five. Now how far things are going to progress after that will depend on many things, the problems we run into, the effort we put into solving these problems. But I think that in the 20-year period, we'll be considering manned flights, certainly to Mars, and probably manned flights around Venus. I don't think we'll be landing on Venus for quite a long time. That's a tough proposition.*

Cronkite: *What about propulsion systems?*

Clarke: *Well, the propulsion systems we have now—that we've demonstrated—can do all these things. They can certainly explore the nearer planets, but within the next five years we should have our first flight nuclear propulsion systems. These have been tested on the ground. We know how to build nuclear-powered rockets and they've been very successfully tested. And the hardware exists. It's a question of getting it in a flyable condition, and when we can do this we will multiply by a factor of two or four the payloads we can put into space.*

Also in this coming ten-year period we've got to see the development of reusable spacecraft. I mean, spectacular and beautiful as the Saturn V is, it's a fantastic way of doing the job. It's like the Queen Elizabeth sailing with three passengers and sinking after the maiden voyage, except that the Saturn V costs more than the Queen Elizabeth. We've got to have spaceships that we can use over and over again as often as we use a conventional airliner. The reusable space transporter is the next thing which we have got to get.

Cronkite: *Arthur, didn't I overhear you say somewhere that these engines, really as remarkable as they are, aren't terribly efficient, that a cupful of kerosene and oxydizer could do the job?*

Clarke: *Well, when you work out the mathematics of the energy involved in carrying a man from the earth surface to the moon, you find you get that amount of energy from about thirty-five dollars' worth of kerosene and liquid oxygen; I think it's about 700 pounds of weight. Now it takes a thousand tons to do the job. So we have a factor of at least a thousand-fold inefficiency theoretically. We will never quite make that improvement. I don't say we're ever going to get to the moon for $35, but we're going to get to the moon for a few thousand dollars. It will be a straightforward commercial operation in another 30–50 years at the most.*

Cronkite: *Do you think that you and I will make a space flight?*

Clarke: *I have every intention of going to the moon before I die. This annoys some of my conservative engineering friends. I'm always annoying them and I often turn out to be right.*

Cronkite: *I hope you're right because I have a feeling that by a little diligence, if you can make it I might get a chance to make it, and I'm dying to go.*

The tension in the studio increased as noon approached. Astronauts Neil Armstrong, Michael Collins and Buzz Aldrin had been in earth orbit just over two hours, and another critical moment in the mission, translunar injection, was due in 16 minutes. The astronauts, out over Australia, would fire the engines of the third stage of the Saturn V to propel them out of earth orbit on the way to the moon. Cronkite and Schirra waited anxiously in the studio, listening to the communications between Mission Control in Houston and the astronauts, and watching a CBS News simulation, originating in New York, of the maneuver.

The word came at 12:16 p.m. EDT.

Neil Armstrong, the commander of Apollo 11 which had reached a velocity of 35,579 feet per second, relayed the word that the firing had been a success:

Armstrong: **Hey Houston, Apollo 11. This Saturn gave us a magnificent ride.**

Houston: Roger, 11, we'll pass that on, and it looks like you are well on your way now.

Armstrong: We have no complaints with any of the three stages on that ride. It was beautiful.

For those involved in CBS News' coverage of the lift-off, the day was drawing to an end. The broadcast would end at 1:15 p.m., and most of the people would travel directly to Melbourne, 30 miles south of the Cape, to catch a chartered flight back to New York. A clean-up crew would stay for a few days to close down the Cape operation before going on to their respective "Lunar Day" assignments in Los Angeles and Houston. The technicians would remain behind to "break down" the studio and control room. Those who would be running the CND Wire would be leaving for Houston where most of the story of the next eight days would originate.

Cronkite was returning to New York for the first time in over four weeks. He had been on vacation for two weeks in mid-June, and then had gone on a week-long, cross-country trip with associate producer Ron Bonn to film Apollo 11 background pieces for the "CBS Evening News with Walter Cronkite." In the six days on the road, he had been in Bethpage, New York; Flagstaff, Arizona; Houston, Texas; Goldstone, Arizona; Downey, California and at former President Lyndon B. Johnson's ranch near Austin, Texas.

He had arrived in Florida on July 6, ten days before lift-off. The ten days were spent filming additional pieces for his evening news broadcast and doing his homework for the coverage of the mission. He had gone over all available material on the mission, querying the research unit about points that seemed unclear or inaccurate. From his studies he had compiled his own source book of information that would be readily accessible when he was on the air.

The massive task of compiling background material on Apollo 11 fell to the research unit under manager Beth Fertik. In at least one case, valuable information came from an unexpected source. The researchers had been having trouble locating a single source of information on the moon's place in literature. CBS News President Richard S. Salant proved to be that surprise source. It seemed that Salant, as a senior at Harvard in 1935, had chosen "The Moon in the Romantic Poets" as the subject of his honors thesis. The thesis won Salant a $500 prize, money he recalls was spent buying his first car.

When the researchers learned of this thesis, they wrote Harvard to see if the original was stored somewhere at the school. It was found, "buried in the bowels of Widener Library," and was sent to New York. The thesis was to prove helpful in putting together broadcast segments on the moon's history and literary significance.

The contribution provided by the thesis 34 years after it was written led Salant to joke that it is best to "never underestimate the utility of a Harvard education."

The flight back to New York was both a period of reflection on the seven-hour broadcast that had just ended, and a look ahead to the 32 hours of Sunday and Monday. Most of the people on the plane had had little or no sleep the night before.

Wussler and Joan Richman and Clarence Cross, the co-producers, had been at the press site with Cronkite the night before for a 10:00 to 11:00 o'clock preview broadcast, and had been back at the press site at 3:30 the next morning to prepare to go on the air at six o'clock. All three were disappointed in the day's coverage. The seven hours had not seemed to have the excitement or continuity they had hoped for, but that was in the past. They turned their attention, as they had done so often for more than two months, to the "Lunar Day" coverage.

CBS News had long been committed to the idea of continuous coverage of the moon landing mission. CBS News Vice President Gordon Manning, who holds the reins on all of CBS News' "hard" news broadcasting, recalls that on the night of the successful Gemini 3 mission in March 1965, he had talked to Wussler about the future of space and television news' role in covering it. Projecting farther into the future than perhaps even the space agency was at that time, Wussler had told Manning that "we'll be on the air all night when men land on the moon."

Just short of four and one-half years later, man was on the verge of taking the historic first step on the lunar surface, and Wussler's and Manning's projected "all night" coverage materialized into 32 hours of continuous coverage. The concept of how to cover Apollo 11 had been arrived at in stages.

For more than six weeks Manning and Wussler had almost daily meetings. It was during this period that the general "theme" of the coverage was decided upon. Manning and Wussler agreed that in reporting the story of man's conquest of the moon equal emphasis should be given to the reactions of people around the world. Once the decision was made to follow this idea throughout the coverage, "Man on the Moon: The Epic Journey of Apollo 11" began to take shape.

Then, on July 4, twelve days before the lift-off, Manning and Wussler had arranged for all the remote producers and correspondents to meet in Houston to discuss "Lunar Day." Both Wussler and Manning agreed that this was a particularly satisfying day, because everyone had a chance to learn what they and Miss Richman and Cross had in mind for the coverage, and the producers and correspondents, all of whom were veteran space reporters, had the opportunity to discuss how various aspects of the coverage could be improved. It was a "hair down" session, and would be looked back upon as one of the most important of the pre-Apollo 11 period.

Manning, Wussler, Miss Richman and Cross became convinced of one thing during this time. While it would be imperative to plan as much of the 32 hours of continuous coverage as possible, it would be necessary to maintain complete flexibility. As Manning put it, "Let's follow that old journalistic maxim: 'Plan for the things you can plan for, so you are ready for the things you can't plan for.' "

They agreed that the most they could hope to do would be to outline the various elements that would make up the broadcast. There were the major domestic remotes—at Houston, where Bruce Morton and producer Bill Headline would cover Mission Control; at the Tomorrowland section of Disneyland, where Heywood Hale Broun and producer Paul Greenberg would cover the reactions of people at a time when all of the world became part of the future envisioned at Tomorrowland; and at Washington's Smithsonian Institution, where Roger Mudd and producer Ed Fouhy would cover national figures in a setting provided by the *Kitty Hawk*, *The Spirit of St. Louis*, the X-15 and John Glenn's Mercury spacecraft.

Correspondents George Herman and John Hart and producer Bernie Boroson would be stationed at Flagstaff, Arizona, where United States Geological Survey astrogeologists had constructed a man-made lunar landing site composed of material similar to the surface they believed Armstrong and Aldrin would find on the moon. Correspondents Bill Stout and Terry Drinkwater and producer Jack Kelly would be at the North American Rockwell plant in Downey, California, where Leo Krupp, the firm's chief Apollo research pilot, would describe the astronauts' activities in a full-scale model of the Command Module. And finally Correspondent Nelson Benton and producer Frank Manitzas would be at Bethpage, New York, the home of the Grumman Corporation, where Scott MacLeod, chief consulting pilot on Lunar Module projects, would simulate Armstrong's and Aldrin's duties during the moon landing. A key element in the Grumman remote was the addition of a 120 by 125-foot moonscape, duplicating as closely as possible the actual landing site, where the moon landing and walk would be simulated.

The world's reaction to man landing on the moon would be a major part of the coverage, and live coverage was scheduled from London, Rome, Paris, Amsterdam, Manila, Tokyo, Belgrade, Bucharest, Mexico City, Montreal, Lima and Buenos Aires. In this country live reaction stories would originate from the International Arrivals Building at New York's Kennedy Airport, from a mobile unit traveling around New York City, and from CBS Television Network affiliated stations in eight cities around the country—Atlanta, Dayton, Hartford, Phoenix, St. Louis, Salt Lake City, Seattle and Wichita.

Arrangements would have to be made for the guests who would appear at various times during the 32 hours. In New York, where most of the guest panels would originate, this aspect of the coverage would be the responsibility of producer Burton Benjamin. Guest interviews also would originate from such widely separated places as Los Angeles,   27

Washington, Houston and London. The guests would have to be informed of when and where they should be, and given some idea of when they would be on the air. You cannot have special guests sitting around for 12 hours waiting to go on television for ten minutes.

Then there was the film bank, the collection of stories filmed especially for Apollo 11. It was the most extensive film bank ever compiled, containing 140 separate pieces representing between eight and ten hours of programming, and including such diverse items as an exclusive interview with former President Johnson, an Orson Welles-narrated science fiction film, four-minute profiles of 15 of the men who had made great contributions to the Apollo program, histories of man in space and lunar exploration, a profile of an 80-year-old bush pilot from Spearfish, South Dakota and a British study of the anatomy of a space suit.

These were some of the elements available for the 32 hours. They could have been incorporated into an hour-by-hour plan, but Wussler, Miss Richman and Cross were determined that the programming would be as loose as possible. The emphasis of the day would be the moon landing and the astronauts. Everything would follow from that.

Cronkite, the producers and the rest of the nearly 1000 people who would be involved in CBS News' "Lunar Day" coverage had three days to prepare for 11 o'clock on Sunday morning. There would be daily coverage of live television transmissions from the spacecraft and of Apollo 11's entrance into lunar orbit on Saturday afternoon. But mostly, the three days would be a period of waiting, as the astronauts, the world and television prepared for the most historic day of their lives.

On the Way
July 17, 18, 19

Thursday was spent checking out the anchor studio in New York, viewing recently edited film, rehearsing simulations and ironing out difficulties at the various remotes in the United States and around the world.

The New York coverage of "Lunar Day" would originate in Studio 41, CBS's largest television studio, which normally houses two daytime serial dramas. In many ways the set designed by Hugh Raisky was as unearthly as the story being covered from there. Cronkite's anchor desk, which was elevated 24 feet above the floor, was set against an artist's conception of the Milky Way. Two 6-foot-in-diameter globes, one a Rand McNally moon globe, the other a Plexiglas conception of what the earth would look like to the astronauts, loomed on either side of the desk.

Sixteen television cameras had been deployed throughout the studio to cover the anchor desk, the status desk from which David Schoumacher would give progress reports on the mission, the interview area and the areas where the intricate models and mock lunar landscapes were set up for simulations. Four cameras were "slave" cameras. Trained continually on clocks that told the elapsed time of the mission and the time left to a major event, and a diagram showing how far the astronauts were into their mission, the cameras were used in a closed circuit operation that fed the information to monitors at the anchor desk.

The simulations and animations were among the more intricate and demanding aspects of the coverage. They were the responsibility of director Joel Banow who would spend another day rehearsing the undocking, working with his technicians to get them to fly the models just as the astronauts were flying the actual spacecrafts.

Banow had been designing and improving the simulations throughout the Apollo coverage. He preferred that simulations be done "live" whenever possible, that is, synchronized with the actual events. His models of the Command and Lunar Modules were motorized and could be moved by remote control to simulate pitch, yaw or roll maneuvers actually being done by the spacecrafts during the mission.

The simulations seen during the insertion into lunar orbit, the undocking and beginning of the descent to the moon's surface, the rendezvous, docking and the return of Armstrong and Aldrin to the Command Module, were done at the moment the actual maneuvers were taking place in space. The motorized models were superimposed over a simulated lunar landscape representing 60 miles across the lunar surface at the moon's equator. It was rolled to give the impression that the spacecrafts were in orbit, with the lunar surface passing under them.

The actual "burns," when the LM's ascent and descent stage engines were being fired, were animated as in the film cartoons. Banow had studied a copy of the flight plan that the Apollo 11 astronauts were carrying into space and had timed the critical points of the mission. If the flight plan called for a two-minute burn, Banow's animations would duplicate it in length, ideally starting and ending as it was taking place in space.

Once the Lunar Module had landed on the moon, Banow planned to use a tiny model of the LM to show its position on a scale model of the landing site. The model, no bigger than a bumblebee, was stored in a bottle and was manipulated with tweezers.

There was one new addition to Banow's bag of tricks. His or its name was "HAL 10,000," a computerized system designed by Banow and Doug Trumbull, who was responsible for most of the special effects in *2001: A Space Odyssey*. CBS News' "HAL" had been named for the famed mutinous computer in that movie. "HAL" was a four by six-foot rear projection screen onto which 12 film projectors could flash images of words and diagrams and had the ability to instantly call up information pertinent to the mission. Thousands of words of space vocabulary and hundreds of diagrams had been stored in the eight 16 mm projectors that were the heart of the system. Each word or image was stored on a single frame of film, and each of the eight projectors held up to 10,000 frames. "HAL" was also able to talk to Cronkite, and was programmed to say "Good evening, Walter," or "I'm looking forward to lunar touchdown."

The only editorial responsibilities of the day had to do with the first scheduled television transmission from the Command Module. The word "scheduled" had to be taken with a grain of salt because the television industry had learned on Apollo 10 that transmission schedules meant little when there were astronauts who also were television enthusiasts. On that flight, with Tom Stafford, Eugene Cernan and John Young on board, everyone learned to expect unscheduled and frequent pictures from space.

On Apollo 11, the first "scheduled" transmission was in fact the third sent to earth by Armstrong and his crew. It was scheduled to begin shortly after 7:30 p.m., and Cronkite, who had just completed his stint on the "CBS Evening News with Walter Cronkite," was in the studio with Wally Schirra. The pictures came up shortly after 7:30, on schedule, and the first thing seen was a view that Neil Armstrong described as "just a little bit more than a half earth." The screen showed the eastern Pacific Ocean, and parts of Alaska, Canada, the United States, Mexico and Central America. Armstrong was also questioned by Houston about cloud formations which seemed to cross Mexico and the northern section of Central America.

Then Mike Collins decided to give CAPCOM Charlie Duke and the world a little different view of itself:

Collins: Well, hold on to your hat. I'm going to turn you upside down.

Capcom: 11, that's a pretty good roll there.

Collins: Oh, I'd say sloppy, Charlie. Let me try that once again.

Capcom: You'll never beat out the Thunderbird. That practice did you some good. It's a real smooth-looking roll out there.

Collins: Ooops.

Capcom: Spoke too soon.

Collins: I'm making myself seasick, Charlie. I'll just put you back right-side-up where you belong.

33

As Houston was asking the astronauts to "see some smiling faces up there," and the astronauts were preparing to take the television camera inside the spacecraft, Cronkite and Schirra chatted about the pictures they had just seen and talked about Charlie Duke's reference to "the Thunderbird."

Cronkite: *Wally, while they're doing that, what were they talking about—the Thunderbirds—when they were doing that roll, turning the camera over?*

Schirra: *It's an old fighter pilot's trick we used with some early airplane pictures. We'd get the effect of a slow roll by turning the camera. And this would give you an effect of doing aerobatics. Of course, they're referring to the Air Force stunt team, the Thunderbirds.*

The astronauts put on a big show for everyone that night. Neil Armstrong stood on his head, Mike Collins disavowed the use of cue cards, telling Houston, "we have no intention of competing with the professionals," and Buzz Aldrin did a few "zero G" push-ups. Viewers were treated to a demonstration of how to make chicken stew when you are traveling at a speed of 4400 feet per second, more than 139,000 miles from earth.

The communications between the astronauts and Houston revealed an interesting fact about television from space. Armstrong trained the camera on the clock on the control panel which records the elapsed time of the mission in hours, minutes and seconds. When Charlie Duke read the time he was seeing on his monitor back to Armstrong, he was told that he was 11 seconds behind the astronauts' actual time in flight. CBS viewers learned that the time differential was due to the delay between the transmission of the signal and the time it reached the home television screen. The reason for the delay was explained by Schirra: "At that distance, the signal should take less than a second to reach us, since it travels about 186,000 miles per second, that is the speed of light. However, the signal goes to Houston, is converted and then sent to us. A good part of the eleven-second delay is due to the time it takes to process the signal through the equipment."

Armstrong ended the transmission with a view of earth receding in the distance, saying "as we pan back out to the distance at which we can see the earth, it's Apollo 11 signing off."

Friday, July 18

Friday was another day spent checking details which seemed miniscule at the time, but, if missed, could become monumental on Sunday and Monday. One piece of good news came from Flagstaff, site of the most remote "remote" of them all.

While the man-made lunar surface at the United States Geological Survey installation in Flagstaff was the prime area to be used in CBS News' coverage, plans also included extensive use of Meteor Crater, which was some 35 miles from the CBS News headquarters. The producer Bernie Boroson had had his headaches. The first problem arose when a dune buggy club had scheduled a rally in the area known as Cinder Lake, where the landing site was located. The dune buggies would not be racing on the landing site, but they would make a great deal of noise, and the sound of roaring engines and the dust they raise would be out of keeping with the stillness and solitude that would greet the astronauts when they stepped onto the lunar surface. After protracted negotiations, the dune buggy club agreed to postpone its rally.

The Department of the Interior also came into the picture when Boroson learned that each morning they salted the clouds over Flagstaff to induce rain. Well aware that it did not rain on the moon, he contacted the local office of the Department of the Interior, which agreed not to salt the clouds until after July 21.

Because high elevation views of the area around Flagstaff were important to the coverage, Boroson had planned to have a camera situated on the top of a nearby mountain. Someone, thinking he was being helpful, plowed the cinder path that ran up the side of the mountain. From then on, any truck trying to make the trip sank into three feet of cinders loosened by the plow.

The lifesaver in this operation was Don McGraw, the engineer-in-charge for the Cape lift-off coverage, who outfitted a helicopter with a portable color television camera. This helicopter had been used by the television pool to telecast live aerial views of the launch site and the surrounding area. As soon as the lift-off coverage ended on Wednesday, McGraw began dismantling the helicopter to get the equipment to Flagstaff. The equipment was packaged and aboard the two private jets late that afternoon, and arrived in Flagstaff that evening.

Word was sent to New York that Flagstaff would be ready on Sunday morning.

The Apollo 11 astronauts had been scheduled to begin their second scheduled television transmission shortly after 7:30 p.m. EDT, Friday night. That schedule was tossed out the window when Armstrong and Aldrin decided to take the camera into the Lunar Module during its initial check-out preparatory to entering lunar orbit the next day.

When Cronkite and Schirra went on the air at 5:50 p.m. EDT, the television screen showed Aldrin in the Lunar Module 203,200 miles from the earth, taking an inventory of the equipment stowed in the spacecraft. Viewers were able to eavesdrop on an interesting exchange between the astronauts and CAPCOM Charlie Duke in Houston:

Capcom: **11, Houston. Buzz, are you still looking for that 90-degree bracket? Over.**

Armstrong: **Yeah, he's still looking for it now.**

Schirra: *A little game of hide and seek here. Apparently one of the components was stowed improperly. When we get the myriad of little gadgets and pieces of equipment that are stowed for launch through the agony of the boosted flight, then try to go find them again, you have what we call a routine C squared, F squared, which is crew compartment fit and function. This is quite an agonizing exercise but you hope all the parts are there so that you can practice finding them, practice assembling them.*

36

Cronkite: *Well, is it possible that a part would be left out? I mean, he's looking for an L-bracket. Is it possible that back on...*

Schirra: *Could be on the beach. There we go.*

Aldrin had found the bracket and was demonstrating how it would be used to steady the movie camera that would film the descent and ascent stages of the mission, and show the astronauts during their walk on the moon. Schirra explained that it would be used to "mount the camera and give it the proper field of view out the window so it doesn't have to be hand-held. We're learning a few tricks from your trade."

Aldrin also gave his huge television audience, which Duke informed him included live coverage in the United States, Japan, Western Europe and much of South America, a look at the space suit and life support equipment he and Armstrong would wear on the moon. When he picked up the helmet to show how the two visors worked, Schirra noted that if he were doing that on earth, he would be in trouble.

Schirra: *There's an interesting facet to this. There's no quality control inspector to chew him out for touching this with his bare hands. He should have gloves on. Of course, we don't have those men along with us at this point.*

Cronkite: *The idea of that is that any amount of oil from your hands that gets on that surface could change the thermodynamics of the material itself.*

Schirra: *Well, it's not really as bad as that. The real problem is that if everyone who touched that left fingerprints on it, it would wear the surface off trying to clean it. And this would destroy its value. And there is the possibility of course that you could pick up flaws from it.*

Cronkite then switched to the CBS News studio at Grumman, where Correspondent Nelson Benton and Scott MacLeod, chief research pilot for Lunar Module projects, were in a full-scale mock-up of the LM. MacLeod told of one item Armstrong and Aldrin had not discovered during their inspection of the LM—a ham salad on rye bread sandwich. The sandwich had been requested by one or both of the astronauts and was put aboard with the blessings of Mission Control.

Cronkite reminded Schirra of a sandwich incident on Gemini 3 in March 1965, in which he was involved with John Young, Gus Grissom's copilot on the flight. That sandwich made it into the spacecraft without Mission Control's blessings.

Cronkite: *You know the kind of food they're putting aboard spacecrafts these days shows how far we've come, Wally, since you kind of got your wrists slapped for slipping a sandwich aboard on an earlier flight.*

Schirra: *Well, actually I didn't get slapped as much as poor John Young did. I had the opportunity, of course, to cater and it was up to John to take it aboard or not. And of course he elected to. At that time we had no programmed meals on that early Gemini flight. Now we're up for over a week. We must have programmed meals. And it's kind of interesting to hear about a disassembled sandwich in flight.*

Cronkite: *What do you mean, you catered for it?*

Schirra: *I obtained the sandwich at one of the restaurants in Cocoa Beach and brought it out to the crew quarters the night before launch.*

Cronkite: *John had asked you to do that?*

Schirra: *He had, yes.*

The "Man on the Moon: The Epic Journey of Apollo 11" special report ended at 6:30 p.m. Schirra left and Cronkite stayed in place to do an Apollo 11 piece for his evening news broadcast. The Apollo 11 astronauts were rapidly drawing to within 40,000 miles of the moon, and in just under 19 hours from that time would fire the engine of their Command Service Module to brake the spacecraft into lunar orbit. Man would take his next step toward the moon.

One day to go. Wussler, Miss Richman, Cross and Banow had been in the office till nearly midnight the night before, and had come in at eight o'clock that Saturday morning to get an early start on the final day of preparations for "Lunar Day."

They concerned themselves with further refining of the schedule of broadcast elements. Some film had to be edited, including the major Orson Welles-narrated science fiction package, but on the whole the packaged material was in good shape. There also were last-minute checks on the simulations and animations to be used during the 32 hours.

There were two special reports that day. The first, from 1:30 to 2:00 p.m. EDT, covered the spacecraft's insertion into lunar orbit, and later in the afternoon there was another television transmission from space.

The chief concern was the technical aspects of the "reaction remotes" that would be used the next day. Wussler had spent a week in Europe at the end of June, lining up the European remotes. Coverage from London, where Correspondent Mike Wallace was stationed as the anchor man for the coverage in Europe, and from Paris, Rome and Amsterdam, had been confirmed weeks before. Wussler had been trying for some time to arrange a live pickup from a Communist country. While in Germany, he learned that there was a possibility that JRT, the Yugoslav television network, would be interested in providing live coverage from Belgrade. JRT would supply five English-speaking correspondents at locations in and around Belgrade. Correspondent William McLaughlin, who normally covers Bonn and Berlin for CBS News, was assigned to the Yugoslav capital.

After President Nixon's around-the-world trip had been announced, another Communist country was added to the list. Diplomatic Correspondent Marvin Kalb, who would be covering part of the President's upcoming journey, would report from Bucharest, Rumania, the next-to-last stop on Mr. Nixon's itinerary.

All of the foreign remotes were confirmed and correspondent and producer assignments had been made. Wussler's "man-in-charge" in Europe was Marshall "Casey" Davidson, who would coordinate the coverage on the continent. London would not only be the European anchor location but would serve as the center for routing all European coverage to the United States. This post was manned by Ralph Paskman and William Small. Morley Safer and producer Daniel Bloom, the London Bureau Chief, would cover activity at the Jodrell Bank observatory in Manchester, England, where Sir Bernard Lovell, director of the observatory and one of the world's leading radio astronomers, would be special consultant to CBS News.

Other correspondent and producer assignments included Peter Kalischer with Norman Gorin in Paris; Winston Burdett with Frank Fitzpatrick and Mario Biasetti in Rome; Robert Schakne with Alan Harper in Mexico City; Daniel Schorr with Robert Chandler in Amsterdam; George Syvertsen with Robert Little and Jeff Gralnick in Tokyo; and Richard Threlkeld with Don Hewitt in Manila. Correspondent Don Webster and Reporter Tony Sargent with bureau chief David Miller were set to report from Vietnam. Their coverage would be videotaped and flown to Manila for transmission via the Pacific satellite to New York. Canadian Broadcasting Corporation Correspondent Sheridan Nelson would report from Montreal, and the coverage from Buenos Aires and Lima would feature English-speaking correspondents from the Argentine and Peruvian television networks.

In addition to the foreign remotes, Manning and Sid Kaufman, the producer who would be in charge of all remotes, had arranged for local reaction coverage by eight CBS Television Network affiliated stations. They included WAGA-TV in Atlanta, Georgia; WHIO-TV in Dayton, Ohio; WTIC-TV in Hartford, Connecticut; KOOL-TV, Phoenix, Arizona; KMOX-TV in St. Louis, Missouri; KSL-TV in Salt Lake City, Utah; KIRO-TV in Seattle, Washington; and KTVH in Wichita, Kansas.

Never before had so many foreign and domestic remotes been set up for the coverage of an event. A separate control room had been set up to handle the feeds coming into New York from the various locations around the world. The major domestic remotes, such as Grumman, Downey, Flagstaff, Disneyland, the International Arrivals Building at Kennedy Airport in New York, the Smithsonian and Houston, would be fed to the control room on lines that would be open at all times. The transmissions from the affiliates would be handled as they came in, with each affiliate getting advance warning that CBS News would be coming to them shortly for a story.

The routing of foreign coverage, all of which was to be supplied live via satellite, had been simplified, although at great inconvenience and at considerably greater cost, by the breakdown of the Atlantic satellite, Intelsat 3, two weeks before. Because the Atlantic satellite had failed, all transmission from Europe would have to be sent around the world via Asia. A complicated system had been arranged whereby the signal from London, Paris, Amsterdam or Rome would be sent to London via telephone lines, then to the south of England for transmission to the satellite hovering over the Indian Ocean. From there the signal would go to Japan for transmission via the Pacific satellite to the west coast of the United States, and via land lines to Kaufman's control room in New York. The west-to-east, round-the-world transmission would mean a two-second delay between the time of origin in say, Amsterdam, and the time it was aired in New York. The system, though complicated, had proved successful earlier in the month during Network coverage of Prince Charles' investiture as Prince of Wales.

The New York studio and the other major domestic remotes—Grumman, the Smithsonian, Houston, North American Rockwell, Disneyland, the International Arrivals Building and the New York Flash Unit—had called for 65 cameras and 334 men. In addition the technical requirements of "Lunar Day" at the affiliate and foreign remotes called for an additional 77 cameras and several hundred technicians. Altogether, 142 television cameras would be in use during the 32 hours, more than twice the number used to cover any other story. (The previous high had been 58 cameras for the 1968 Republican National Convention in Miami Beach, Florida.)

Logistical problems were not the exclusive property of those directly involved in the coverage of Apollo 11. Frank Smith, Vice President of Sales for the CBS Television Network, was having problems of his own. Never before had he been faced with the prospect of selling sponsorship for a news event that included 32 hours of continuous broadcasting, and called for scrapping more than 15 hours of scheduled Network programs.

Although everyone at CBS had accepted the fact that the coverage of Apollo 11 would result in a financial loss, Smith was responsible for keeping that loss to a minimum.

Smith admits that the only nervous moment he had came two weeks prior to lift-off when Western Electric, which had assumed full sponsorship of the earlier Apollo missions, and which Smith had expected to sponsor the entire Apollo 11 coverage, decided to sponsor only one-third. "There was a time," Smith says, "when I wondered whether we would find sponsors to take up the slack on such short notice."

Then International Paper Company, making its first network television purchase, bought one-third of the coverage, and Kellogg and General Foods each picked up one-sixth of the package. Apollo 11 would be fully sponsored. Even though these sales were at the multi-million-dollar level, the return would fall nearly two and one-half million dollars short of matching the costs incurred in bringing the event to the American people.

Providing live coverage of Apollo 11 to Alaska posed a unique challenge. The three stations in Anchorage, affiliated with ABC, CBS and NBC, had never received live television from the mainland United States. Viewers in Alaska previously were dependent on videotape or film, but the significance of the moon landing and walk had led Alaska Senators Gravel and Stevens and Congressman Pollock to make every effort to have Apollo 11 broadcast live in their state via satellite.

The final arrangements were made, with the cooperation of the U.S. Army, 72 hours before the launching. The Army flew a mobile television ground station to Anchorage. CBS News' coverage was chosen by lot, and all three stations in Anchorage agreed to broadcast CBS News' coverage of the mission. The signal sent to Alaska was that of WCBS-TV in New York, transmitted from Fort Monmouth in New Jersey via a satellite to the receiver in Anchorage. The only parts of the broadcast the Army would not send were the commercials, which were deleted at Fort Monmouth.

Canada represented another special situation. The Canadian Broadcasting Corporation fills a part of its broadcast schedule with programs produced in the United States. The great Canadian interest in Apollo 11 led the CBC to decide to carry the full Apollo 11 coverage of one of the American networks. They chose CBS News and applied for permission to have the coverage sponsored. And for the first time the Canadian Parliament agreed.

At 1:25 p.m. EDT, Cronkite and Schirra were in place in the studio for a 1:30 to 2:00 p.m. special report covering the critical lunar orbit insertion phase of the mission. This had been one of the more dramatic moments on the Apollo 8 and Apollo 10 missions. There was a 23-minute wait from the time the astronauts were to fire the service propulsion system engine that would brake them into lunar orbit, to the notification that the firing had been a success. If the engine failed to fire properly, the spacecraft would whip around the moon on its way back to earth. The mission would have failed.

Cronkite: *We don't know if all is going well with the Apollo 11 because it is behind the moon and out of contact with earth for the first time. Eight minutes ago they fired their large service propulsion system engine to go into orbit around the moon. We'll know about that in the next 15 minutes or so. That's when they come around to the near side of the moon and again acquire contact with the earth and they can report. We hope that they are successfully in orbit around the moon and that the rest of this historic mission can go as well as the first three days.*

The final exchange between Houston and Apollo 11 was simply, "Apollo 11, this is Houston. All your systems are looking good. Going around the corner and we'll see you on the other side." From the spacecraft came a terse, "Roger."

The tension felt by Cronkite and Schirra, and by those working 43 and watching in the control room at CBS, was matched by the reaction in Mission Control.

Houston: **There are a few conversations taking place here in the control room, but not very many. Most of the people are waiting quietly, watching and listening. Not talking.**

Cronkite: *That's the word from Houston. It is quiet there. It is quiet around the world as the world waits to see if Apollo 11 is in a successful moon orbit.*

The final seven minutes before Acquisition of Signal, the positive sign that the astronauts had achieved a lunar orbit, were mostly silent. The countdown for Acquisition of Signal (AOS) reached the last 30 seconds, and Cronkite described the wait for word from Neil Armstrong, as a CBS News simulation of the moment flashed on the television screen.

Cronkite: *It's 1:47 and sometime perhaps in the next 20 seconds, if all is exactly on schedule, we should be getting that signal from Apollo 11. It should be just coming around toward the near side of the moon now and if the firing of the service propulsion system engine was exact —and there's every reason to believe now that it was—the new orbit will be...*

Apollo 11: ...[indistinct signal from spacecraft]...

Cronkite: *The signal! And right on schedule! How about that! Almost to the second as predicted in the flight plan! It came around the moon, a good data signal. And that may not be a voice transmission yet. That could be simply acquisition of the telemetry signals, the thousands of bits of information that are transferred back every second from the spacecraft systems. That roar you hear in the background is Houston tuned up for the first words to be heard as soon as we get something from the astronauts.*

The waiting ended.

Houston: **Apollo 11, Apollo 11, this is Houston. How do you read?**

Collins: **Read you loud and clear, Houston.**

Apollo 11 was in lunar orbit, and the spacecraft was making its first pass over the southwestern section of the Sea of Tranquility, where the Lunar Module would land the next day.

Armstrong: **Apollo 11. We're getting this first view of the landing approach. This time we are going over the Taruntius Crater and the pictures and maps brought back by Apollos 8 and 10 give us a very good preview of what to look at here. It looks very much like the pictures, but like the difference between watching a real football game and watching it on TV—no substitute for actually being there.**

Capcom: **Roger. We concur and we surely wish we could see it firsthand.**

The thrill the astronauts were experiencing in seeing the landing site was shared by television viewers later that Saturday afternoon.

The scheduled television transmission from Apollo 11 as it circled the moon began at 3:50 p.m. EDT. Correspondent David Schoumacher was in the studio with Newell Trask, the United States Geological Survey astrogeologist who had coordinated the mapping of the area targeted as the landing site for Armstrong and Aldrin. The television transmission consisted of pictures taken as the spacecraft orbited nearly 100 miles above the lunar surface. The landing site, which was near the terminator, the point where the astronauts would pass from light into darkness, was clearly visible.

As they crossed the terminator, they trained the television camera back through the window for a last look before sign-off. The Apollo 11 astronauts had just passed the 24-hour point in the countdown for the lunar landing, and in an hour would fire the service propulsion system engine to lower them into a circular parking orbit 69 miles above the moon. Armstrong and Aldrin would go back into the LM to re-check its systems; then the astronauts would rest.

Tomorrow was the big day.

Man on the Moon
July 20, 21

The voice is that of CBS News Correspondent Charles Kuralt. It is 11:00 a.m. EDT, Sunday, July 20:

In the beginning God created the heaven and the earth. And the earth was without form and void. And darkness was on the face of the deep. Some five billion years ago, whirling and condensing in that darkness, was a cloud of inter-stellar hydrogen, four hundred degrees below zero, eight million miles from end to end. This was our solar system, waiting to be born.

A hundred million years passed. And God said "Let there be light." And there was light at the center of that whirling cloud as a protostar began to form, its gravitational pull attracting larger and larger mass —rotating faster and faster. And in that condensation and heat the sun was born—in fire out of cold—ever smaller and ever more brilliant, ringed with those satellites that were to be its planets. Two protoplanets —the earth and the moon—now separate gaseous eddies mutually trapped in their gravitational pull, moving in tandem orbit around the sun and growing more dense. Through space and time the increasing gravitation of the system drew in more and more debris, the heavier elements converging in those burning clouds to form molten cores at their centers.

And more millions of years passed. The earth and the moon drew slowly apart, their rotation about each other gradually decreasing in the expanding universe. By now, most of the dust in the solar system had disappeared, either swept up by the planets, or condensed into solid particles of varying size—meteoroids that roamed in eccentric orbits through the vastness of space. Millions of these wandering objects showered onto these new planets, disappearing into their molten surfaces, leaving no trace of their impact.

And time passed. The moon, being smaller than the earth, began to cool first and its outer crust to harden. But this hardening was interrupted from time to time by gigantic asteroids colliding with the moon and breaking through the crust, opening fissures that released seas of molten lava from within, flooding vast areas of the lunar landscape. As the molten seas flowed outward and over the torn surface they

49

formed the plains or mares. Impact debris projecting above the flows became the mountains of the moon.

And, with time, the crust would cool again and as it cooled, fissures opened, wrinkles and ridges formed. And volcanic pressures under that surface raised great domes that would later collapse leaving crater-like formations when the eruptions subsided. Most of the lunar craters though were probably being formed by bombardments that had been going on from the moon's earliest beginnings: a continuous shower of meteoroids upon the lunar surface, some of the larger ones breaking through the crust and releasing lava from below.

Millenium upon millenium, this cosmic rain persisted. Through billions of years, meteorite debris has been collecting upon the surface of the moon. Even the level areas caused by the lava flows are now deeply buried. Nearly every square foot of this outer layer is pitted with impact craters of its own. Seven hundred thousand years ago came perhaps the most recent large alteration of the lunar surface. A gigantic meteoroid struck the southern mountainous region, creating the Crater Tycho with an explosive force that scattered material in a radiating pattern nearly halfway around the moon. Some of this material was driven out into space and into the earth's gravitational pull, penetrating our atmosphere and falling to earth among the life on our planet at that time.

As we look back through space and time at the origin of the moon, we may be in part contemplating our own beginnings. Were it not for the moon and its effect on the creation of the earth, man might not be on this planet at all, gazing into that luminescent face on countless nights, or be reaching out for it on this day.

Cronkite had been in the studio since 9:30 a.m. He had done some last-minute boning up on the moon landing, thumbing through the ten separate notebooks he uses for reference during an Apollo mission. Each notebook provided information on various aspects of the flight. One, put together by Cronkite in the days before lift-off, was a compilation of pertinent facts about Apollo 11 and about the space program

50

and its people dating back to the late 1950s. This notebook would never be very far from his side during the next 32 hours.

Cronkite never got the eight to ten hours of sleep he had hoped for during the preceding night. The excitement and anticipation had been just too much. When he arrived, he had found that his experience was not uncommon. No one had had much sleep the night before.

Cronkite first went on the air at 10:00 a.m. for a five-minute progress report on the mission. This was followed by "Nearer To Thee," a religious broadcast produced by Pamela Ilott, Director of Religious Broadcasts for CBS News. The broadcast featured a discussion of the religious aspects of Apollo 11 by theologian Theodore A. Gill, sculptor Richard Lippold, writer and space executive Edward G. Lindaman and physicist William Davidon.

Shortly before eleven o'clock, everyone was in place in the control room. Joel Banow and his associate director Richard Knox sat at the console directly facing the decks of monitors that would bring them pictures from the key cameras in the studio and the major remotes. Wussler stood in between the directors' console and a rear deck where his communications set-up was located. He would be concerned not only with what was on the air "now," but with what should come up "next." Seated directly behind was CBS News Vice President Gordon Manning, who would have overall editorial supervision of the broadcast. At his side were Clarence Cross and Wussler's administrative assistant Mary Kane, who would help with phone communications and relay instructions to remote producers. Wussler's key direct phone lines were to Manning and Joan Richman, who was in her anchor producer's spot at Cronkite's right hand. Her desk, some three feet below Cronkite's and out of camera range, had a set of television monitors duplicating those in front of Cronkite. All communications to Cronkite were directed through her. Beth Fertik, the research manager, located at floor level directly under the anchor desk, was responsible for answering questions from Cronkite.

In the editorial area producer Sanford Socolow and several writers were putting together status reports on the mission and compiling brief stories of interest gathered from the Associated Press, United Press International and CBS News' own CND Wire, which was filing stories from Houston.

The first hour of "Man on the Moon: The Epic Journey of Apollo 11" was devoted to setting the scene for the coverage to follow. In his introduction, Cronkite referred to the day's events as "a giant step," little realizing that before the day was out similar words, "a giant leap," would be on the lips of millions of people around the world.

A "whip-around" to the major domestic remotes was one of the first orders of business. The coverage switched to the Grumman plant at Bethpage, New York, where Correspondent Nelson Benton in CBS News' full-scale model of the Lunar Module talked to Scott MacLeod about the reactions of the people at Grumman on the day when the real LM would be receiving its ultimate test. MacLeod, who had devoted years to testing the spacecraft's systems, called it "the most exciting day I've experienced."

From Houston, Correspondent Bruce Morton described the mood in Mission Control as the tension built during the few hours remaining before lunar touchdown. Reporters Marya McLaughlin and Jed Duvall, also stationed in Houston, reported on the activities of the astronauts' wives and the NASA leaders.

From the North American Rockwell plant in Downey, California, Correspondents Terry Drinkwater and Bill Stout were the next to report. Drinkwater was in the model of the Command Module with test engineer Leo Krupp, who described the busy and lonely day in store for Mike Collins. Stout, along with special consultant Dr. Krafft Ehricke, North American Rockwell's Senior Scientific Advisor for Advanced Systems, was stationed in a model of the solar system—a new addition to Apollo coverage that had been developed by CBS News and North American Rockwell. It was a 40-foot-in-diameter, motorized walk-through model, the only fully operational one of its kind in the world. Jack Kelly, the producer of the remote, had had second thoughts about using the solar system as a studio. He was haunted by the thought that Jupiter, 27 inches in diameter, would come spinning around the sun and knock down one of the guests.

The whip-around continued with brief visits to Flagstaff, where Correspondents George Herman and John Hart described activities at the United States Geological Survey installation and at the simulated lunar surface in Cinder Lake; to Van Cortlandt Park in New York City, where Correspondent Bill Plante would cover people's reaction in the big city; to the International Arrivals Building at John F. Kennedy Airport in New York, where Correspondent Joseph Benti would interview arriving and departing passengers who could watch the moon landing on a huge Eidophor screen CBS News had erected; and to the Smithsonian Institution, where Correspondent Roger Mudd stood with tourists in the shadows of the Wright Brothers' *Kitty Hawk* and Lindbergh's *The Spirit of St. Louis*, reminders of other great days in aviation history.

Noting that "something about the astronauts has made us all believers," Cronkite concluded a summary of the mission to that point with a discussion of the superior quality of the television pictures received from space during the past three days. As good as the color television had been on Apollo 10, the pictures received from Apollo 11 had been little short of extraordinary. Dr. Peter Goldmark, President of CBS Laboratories and the man who developed the field sequential color system aboard the spacecraft, talked with Cronkite about the system and how it came to be part of America's manned space program.

Cronkite: *Dr. Goldmark, how is it that a tiny camera out there in space can send back a better picture from a quarter-of-a-million miles away than I can get up on 84th Street in New York City from a huge studio camera?*

Goldmark: *It started this January. NASA asked us to come down to Houston and demonstrate our small medical color camera. Now this medical camera has been used at the Army's Fitzsimons Hospital in Denver for examination of internal organs in the human body. It is tremendously sensitive, requires very low power and it weighs very little. When Colonel Stafford saw the camera he said, "This is what we've got to have." NASA then examined the possibility of converting Westinghouse's black and white camera to this system. At the same*

53

time we gave NASA the information necessary to convert the signals from this camera for broadcasting. This is how it came about. It is interesting that from the small details of the human body to the vast spaces of the moon this camera represents a link.

Cronkite: *Why is it that we don't have color on the landing camera?*

Goldmark: *We almost had it, but this camera—the color part—was really sort of an afterthought and there wasn't enough time. The camera for landing is vacuum-packed, so to say, and I understand, as you said before, that on future missions they will try to use a color camera as well.*

The foreign remotes began to check in with New York. It was noon in New York, and 5:00 p.m. in Manchester, England, where Correspondent Morley Safer and Sir Bernard Lovell, Director of the Jodrell Bank observatory, appeared on the screen. The radio telescope tracking Apollo 11 was not the only topic of discussion in Manchester that Sunday afternoon, because another radio telescope was following the orbital path of Luna 15, the Russian spacecraft that reportedly was to make a soft landing on the moon, pick up lunar material and return to earth with it. Lovell disclosed that the two spacecrafts' orbits would cross twice. It appeared that Apollo 11 was not the only spacecraft the world would have to keep its eyes on that day.

It was late Sunday evening in Tokyo, when Correspondent George Syvertsen filed his first report. Most people were at home, and those who tried to escape the city's heat were at the seashore. TBS, the Japanese television network working with CBS News, had erected a full-scale model of the Lunar Module in the courtyard of Tokyo's tallest building, and was widely promoting its coverage of Apollo 11, which was scheduled to be continuous from the time of lunar touchdown through the end of the moon walk. TBS was making extensive use of the CBS News simulations, which had been given to them in return for CBS News' right to use their facilities and cameras. Syvertsen's report had one ironic note. As he was reporting, 3500 Japanese students were in the city's streets, demonstrating against the United States.

As the time of the lunar touchdown drew nearer, Cronkite and Eric Sevareid reflected upon what the moment would mean to the world's people.

Cronkite: *Just a little over an hour and a half from now the Lunar Module separates from the Command Module and the moon landing exercise begins. So with things progressing as well as they have for four days, this era of mankind, a billion years of it, of man on this planet, should end in four hours, six minutes and 28 seconds from now. And, Eric Sevareid, I'm not sure after all of this that I'm really ready for it.*

Sevareid: *I ran into a housewife the other day who said it's the greatest thing since sliced bread. What you're saying is very true, Walter. The reaction of human beings on earth to this is just as fascinating as the behavior of that machinery and those men up there. I get the impression that the most powerful force on earth is not this rapid propulsion power, but human habit. This may be finished. It may be a new era. It may be the beginning of a new stage in the evolution of the human species. I don't know. None of us know. But tonight, when it's all over, and tomorrow, every man on earth is still going to put his pants on one leg at a time, and brush his teeth the same way, and conduct arguments with his wife the same way.*

But one of the strange things about all this is the assumption of success. As though this were already done. We're all talking that way. We're talking about the international law that may be involved. We're talking about going to Mars and what that might cost, we're talking about the possibility of the infection of earth, or the infection or pollution of the moon itself. Everything's discounted already. I think this is what modern communications does in a way. We chew it all up in advance so it won't be a surprise. And if there are surprises, it will come as a great shock. It wouldn't have been a great shock to people to have learned that Columbus foundered off Cuba or somewhere, and didn't come back. They rather expected it. We don't expect it in this case. In spite of the fact that billions of brain cells are trying to coordinate here millions of moving mechanical-electrical parts.

Now this is the staggering thing to me—this assumption that all 55 *that man can imagine, he will do and successfully. So, really, it isn't like Columbus and things of that kind. It's not discovery. The discoveries come later in the laboratories.*

One of the three men who had been as close to the moon as any man had been before, Apollo 10 commander Tom Stafford, was waiting and watching in the CBS News studio in Houston with Correspondent Bruce Morton. Stafford had taken his LM down to 47,000 feet above the lunar surface in a maneuver duplicating the one that would be attempted later that afternoon by Armstrong and Aldrin. Stafford described what it would be like for the Apollo 11 astronauts during the final minutes of the descent to the lunar surface.

Morton: *Is there a physical jolt when you fire the engine to brake yourself?*

Stafford: *Well, there's an acceleration, just like when you brake into lunar orbit. Of course, the acceleration is very low, because we had a 10,000-pound thrust engine, and the vehicle to start with weighs 32,000.*

Morton: *How does it feel? I mean, is it like suddenly speeding up in a plane or a car?*

Stafford: *Well, it's like a fast acceleration in a car. Of course, in the Lunar Module, we stand up. We don't have any seats. So Neil and Buzz will have two restraint cables attached to their pressure suits. And they'll have their hands on the control, and they'll do an exact simulation and they will make the one burn. In fact they'll make two burns on the descent engine. You can feel the acceleration in your feet. Where they'll really have acceleration is on the ascent stage. It's a lightweight vehicle, and as the mass goes down, the thrust constantly remains and you get a lot of acceleration.*

Morton: *On Apollo 10, Colonel, you were four or five miles south of the landing site somewhere. How accurate do you think 11 is likely to be?*

Stafford: *Well, again this is the way that Apollo 10 fits into Apollo 11. We were targeted right for the center of the landing site. But the lunar mass and its distribution actually pulled us five miles south of the landing site. So what they have done with Apollo 11 is target it to a "false south" site, five miles north. And then the lunar mass will take it and actually put it right down in the center of the landing site.*

Morton: *They ought to come pretty close then?*

Stafford: *They should be right on the money.*

Morton: *If you were Neil Armstrong, Colonel, would you wait those nine hours or so before you got out of the LM and started to walk?*

Stafford: *Well, the flight plan is flexible, and it's up to the commander, and many other people, and also coordinated back with Mission Control here in Houston. Again, we passed on to them how we felt at the time we passed over the site. It's a fairly long day, and this is why we put in the rest period and the meal period. Naturally, the first thing they're going to do is check over the LM for two hours and make sure that they don't have to go back on an ascent. But if Neil feels in great shape, certainly they can start it early.*

Wally Schirra was with Cronkite at the anchor desk, and the two of them joined in the conversation.

Cronkite: *Tom, this is Walter Cronkite, in New York. Wally Schirra is sitting here alongside of me, your companion on Gemini 7, in which you guys really paved the way for this by achieving the first rendezvous in space. If that hadn't taken place we wouldn't be where we are today.*

Stafford: *Right, Walter.*

Schirra: *People have asked me what I thought would be the most difficult part of the mission. I'm not about to talk about stepping on the moon. I think that the landing itself is the difficult part. We've always trained to avoid surprises and this is the one area there might be one. But I can't think of any other area that would cause us any trouble.*

Stafford: *That's right, Wally. I think that the biggest risk factor is the touchdown. If we end up in good shape there, there's far less risk to work out in the vacuum. We've done that a lot before. We know where we are. Everything's been tested in the chamber. To me the whole risk factor is that final 500 feet down.*

Cronkite: *How far over can the lunar lander be canted in that landing? Suppose it lands on one side of a crater? I've heard something about 10 degrees? Is that the maximum it can be tilted over and still make a successful takeoff?*

Stafford: *Well, Walter, the ultra specification for the lunar lander is to land on an angle of about 10 degrees. And then we have crushable aluminum honeycombs in the struts that can take up some of the inclination. But I think he should be able to put it down within 10 degrees.*

One of the world's more interesting places to reflect upon the significance of Sunday, July 20, was at the "Man and His World" exhibition erected on the site of Expo '67 in Montreal, Canada. Canadian Broadcasting Corporation Correspondent Sheridan Nelson reported on the mood at the fairgrounds. The exhibition arranged to stay open all night so that the thousands of visitors could watch Neil Armstrong's first footstep on the lunar surface on a huge Eidophor screen that had been placed on the grounds by CBC. The young people at the exhibition seemed to express what most people felt, excitement over the upcoming landing and concern for the safety of the astronauts.

CBS and CBC had more in common during this coverage than sharing a correspondent. Throughout the coverage, three men from CBC, working in shifts, were stationed in the New York control room behind Wussler, to inform their network of what was coming next and when to expect commercial interruptions.

Excitement and anticipation were the key words in America's reaction to the lunar landing. One after another, the affiliate stations reported—Ray Moore of WAGA-TV, Atlanta; Don Wayne of WHIO-TV, Dayton; Bill Haskell of WTIC-TV, Hartford; Bob Davies of KOOL-

TV, Phoenix; Barry Serafin of KMOX-TV, St. Louis; Dick Norris and Ted Capener of KSL-TV, Salt Lake City; Cliff Curke of KIRO-TV, Seattle; and Ollie Thompson of KTVH, Wichita—telling what the people in their cities were doing on the day men landed on the moon and describing their reactions.

In California, Heywood Hale Broun was at Disneyland.

Broun: *In a sense, Disneyland, with its kind of tremendous and well-organized gaiety, represents the intense optimism which is American. An optimism which gets very upset, frazzled when things go wrong; but when things go right, is so happy, so well-organized, which is what has made us get to the moon, so well and so quickly. And here is the place from which it is perhaps best to watch a moon landing, because they have rockets, the kind of small rockets the children all like. The rockets which don't go to the moon, but go in a vast circle into imagination, which is further away than any moon, or any star or any planet. They go as far as the minds of the children will travel in them.*

And it is imagination which has taken us across the empty seas of space, to now, the back side of the moon. Mickey and Pluto, and Goofy and the Disney family have somehow all become space-conscious, because it is a part of America at this moment to be space-conscious. And so, to those who are on the rides today, there is a feeling that somehow, they are also spinning in an orbit, as spinning in an orbit are our three astronauts, now at that part of the moon which we never used to see. And which used to be, perhaps, the only really wonderful mystery we had.

This day was proving just how thin a line exists between imagination and reality. Correspondent Harry Reasoner was sitting in the New York studio with two men who had long before proved that their imaginations were as fertile as any in the world. The meeting between Arthur C. Clarke, whose *2001: A Space Odyssey* was well on its way to becoming the world's best-known science fiction book, and Kurt Vonnegut, Jr., the "black humorist" whose latest novel, *Slaughterhouse-Five,* has a science fiction theme, was not accidental. Clarke, one of the foremost proponents of America's manned space program, had been taken to task in a recent magazine article by Vonnegut, who thought there were greater priorities here on earth. Gordon Manning had persuaded Vonnegut to come to the studio to discuss some of the differences he had with Clarke.

Reasoner: *There have been a number of people who have questioned the vast expenditures, especially in the last couple of years. I think it's true, probably, that today the critics of the space program are taking a day of silence. But it would be foolish to say that that kind of criticism doesn't exist among a lot of people of good will. You must have run into it a good deal, haven't you?*

Clarke: *Yes, I'm always running into it. Of course, anything one does could be done better in some other way. If you build a school it means you can't build a hospital. There's always a question of priorities. But I think in the long run the money that's been put into the space program is one of the best investments this country has ever made. Because you're going to get back a so-called "spin-off" which NASA's always talking about, which is important, but not as important as the real thing. This is a down payment on the future of mankind. It's as simple as that.*

Vonnegut: *Well, I've been interested in this "spin-off" for a long time. I've been led to believe that the ball on my ball-point pen and Teflon were spin-offs. And now I find out that Teflon was a spin-off from World War II. So that only leaves the point on my ball-point pen.*

Clarke: *I think the ball-point pen was around before the space program. But the real spin-offs are going to be largely in the next decade or so. Many of them haven't really come into use. But the spin-off is going to be more knowledge rather than hardware. The ability to do new things which we weren't able to do before. Because the space program has used such an enormous technology, it's going to really revolutionize life and make it much easier all over the world.*

60

At 1:26 p.m. Cronkite reported that Sir Bernard Lovell had confirmed that Luna 15 had changed its orbital path to one similar to that Apollo 11 would take as it made its descent to the lunar surface. The possibility grew that it was preparing to land or was on a close reconnaissance mission.

At 1:30 p.m. Correspondent Mike Wallace, acting as anchor man in London, came through with Europe's reactions.

Wallace: *Actually Europe has all the best of it so far as television air time is concerned. Londoners, for example, will be settling down by their tellys about six o'clock this Sunday evening and they will have a full night of space spectacular ahead of them in prime time. The landing itself is scheduled for shortly after nine o'clock, London Time. And then, those who can, or want to, can take their customary sleep, and arise at the customary time, in order to witness the descent of Neil Armstrong from the "Eagle" to the surface of the moon. Or, should the time of the moon walk be advanced, BBC Radio has arranged a Moon Alarm for those who care to leave their radios on all night. At the sound of that alarm, Britishers will gather again, in front of their television sets, for an historic Monday, translated "Moonday" hereabouts in many of the newspapers.*

The traditionally stolid British seem to be as captivated by the fact and implications of the flight of Apollo 11 as we Americans. One could truly follow the tension on the streets of London this morning and the newspapers are full of the story, as you can see by the headlines in this assortment. From West Germany and France and Israel and Spain, from Italy, and of course, from England, the magazines and papers have headlined the story, with special supplements and counter stories, all this past week. One local observer says that no one man in memory has received this kind of European coverage unless it was the assassination of John Kennedy.

In Paris, Correspondent Peter Kalischer reported that "as everywhere else, the French press is dominated by Apollo 11, and what if

Armstrong should encounter living beings." The Parisians themselves weren't in a great hurry to get excited about the event. Kalischer described their mood as "one of relaxed anticipation." Americans visiting Paris would have the opportunity to see the landing and walk. The U.S. Embassy had erected large television screens on Avenue Franklin D. Roosevelt for their convenience. It was mid-July, and Paris was a tourist city. The Parisians were on vacation.

The Netherlands was a different story. Correspondent Daniel Schorr, in Amsterdam, reported that the country's NOS television network had planned 30 hours of continuous coverage of Apollo 11, "a massive effort for a network that normally broadcasts only weekends and evenings." Schorr reported that "the audience is estimated at 80 percent of the 12 million population of this country, by far the greatest audience that any Dutch television program has ever had."

But the massive television coverage was not the only sign of interest in Apollo 11.

Schorr: *There are other evidences of Dutch excitement about the moon landing. Gas stations in Holland are distributing moon maps, instead of road maps. A leading newspaper has started a contest among its readers to select a new name for the moon. And the Post Office has launched an investigation, because somebody forged his own commemorative stamp of the lunar landing and put it on a letter that was cancelled by an unsuspecting postal clerk.*

Never before had the Communist world shown so much interest in an American space adventure. Correspondent Marvin Kalb was in Bucharest, Rumania, to cover the reaction to Apollo 11 and the upcoming visit of President Nixon.

Kalb: *Here in Bucharest, the Rumanians have become "space bugs" overnight. It's not every day that man attempts to land on the moon. Not every day, either, that a man named Richard Nixon attempts to land in Bucharest on a Presidential visit. Neither has ever happened before. And that naturally accounts for the intense interest which Rumanians have shown in the magnificent adventure called Apollo 11.*

This modest space exhibit is unique. It is inside the courtyard of the American Embassy, and by local law, Rumanian citizens are not allowed to set foot on Embassy grounds. But the Communist Party dictatorship of this maverick East European country has made an exception in this case, pleasing to look the other way, in deference to the event and the upcoming Presidential trip.

It has been a great topic of conversation all week. The launch was seen live on Rumanian television. And the Rumanian press has carried reports from Houston, Washington, Cape Kennedy and the space capsule itself in great detail. The average Rumanian thinks of Apollo 11 a little bit as his own personal adventure—his link to the marvels of the West. Another reason for his feeling good about his increasing independence in the East. Apollo 11, they keep telling the visitor, is for everyone, for all mankind.

The reactions in Rome were widely varied. Gerald Miller was at the Spanish Steps, described by Wallace as "the gathering place this summer, as in summers past, for the young tourists, American and otherwise. Here," he said, "the drop-out generation rubs shoulders with the affluent young from around the world."

Miller's questions drew reactions ranging from "I think sometimes it would be exciting to go there," to "I don't think about it." The young people in Rome were not very excited about the prospect of man on the moon.

Correspondent Winston Burdett was at Pope Paul VI's summer residence in Castel Gandolfo, Italy. Large crowds had gathered in the courtyard of the Pontiff's summer home, hoping to see him once again that evening. The Pope had spent part of his day watching the moon through a large telescope, and had spoken of Apollo 11 at Mass that morning.

Burdett: *His Holiness the Pope says this is a historic day for mankind. That it is true that tonight two men will set foot on the moon. He calls it an extraordinary and astounding achievement that testifies to man's ingenuity, his courage, his fantastic progress.*

The first critical moment of the day in space was coming up. Collins was alone in the Command Module; and Armstrong and Aldrin were in the Lunar Module preparing for the undocking, the next step toward the moon landing.

The separation would take place seconds after the spacecraft came around the near side of the moon. Collins would fire the service propulsion system engine to pull away from the Lunar Module, and Armstrong and Aldrin would be on their own.

Capcom: Hello, Eagle, Houston. We're standing by. Over.
Eagle, Houston. We see you on the steerable. Over.

Cronkite: *That call to Eagle is to the Lunar Module. The Command Module's call-word is Columbia. Our simulation shows a sophisticated maneuver at this time, as Mike Collins in the Command Module takes a good look at the Lunar Module, checks it out by visual observation. He advises the crew of the Lunar Module, Armstrong and Aldrin, that they look good, and advises the ground of that as well.*

Finally, Eagle answered and the anxious moments were over.

Eagle: Roger. Eagle. Stand by.

Capcom: Roger. Eagle. How does it look?

Armstrong's happy voice cut through the 242,000 miles of space to earth: "The Eagle has wings." And Eagle indeed did have wings, as Armstrong pulled the LM away from Mike Collins in the Command Module. Then Collins prepared to ignite Columbia's engine for the final separation maneuver.

Columbia: I think you've got a fine looking flying machine there, Eagle, despite the fact you're upside down.

Eagle: Somebody's upside down.

Columbia: Okay Eagle, one minute to T. You guys take care.

Eagle: **See you later.**

Columbia's engine fired, and the two were separated. Eagle was on its way to the moon.

Eagle: **Going right down U.S. 1, Mike.**

The undocking and separation had been a success. The CBS News control room was quiet and happy. Banow's simulation had been almost to the second, and as the broadcast neared the completion of its first two hours everything had gone as planned.

Despite the enormous interest in what was taking place at this moment, not a single viewer in the vast television audience apparently noticed the subtle change in Wally Schirra's appearance in the minutes after the undocking. Prior to the maneuver, Schirra had been wearing a lapel pin showing a docked Command Module and LM. When he next appeared on camera, after Eagle had pulled away from Columbia, he was wearing two pins. As the coverage continued, it developed that Schirra had lapel pins depicting every major point of the mission, even to the extent of having two little space men for the period of the walk on the moon's surface.

Mike Collins was a lonely man at that moment. As he saw Eagle sweep off below him carrying Armstrong and Aldrin to a place no man had been before, he must have contemplated what the next 27 hours would be like. Cronkite went to Terry Drinkwater and test engineer Leo Krupp at North American Rockwell for an explanation of what Collins would be doing alone in the Command Module. Krupp described Collins' main duty as being "prepared to rescue the LM crew if anything goes wrong." Collins could fly Columbia down within 50,000 feet of the lunar surface in a rescue attempt, if the need arose, but Krupp warned that he would only have from three and one-half to eleven hours to complete the rescue, depending upon when the astronauts left the lunar surface. Krupp expressed considerable doubt that such a rescue attempt would be necessary. Mainly, Collins would be waiting for 5:32 p.m. EDT, the next day, when Columbia and Eagle would once again be one.

The decision that led to this moment was made eight years before by President John F. Kennedy. Dr. John Logsdon, assistant professor of politics at Catholic University and an expert on the decision-making that led to Apollo 11, was with Roger Mudd at the Smithsonian Institution.

Mudd: *Most Americans have the idea, I think, that the decision was sort of a romantic decision. But there were really a lot of "cold war" political factors, weren't there?*

Logsdon: *It was far from a romantic decision. It was a very cold-blooded calculation that the honor, power and prestige of the United States required that we be first in space. Kennedy calculated that space was a symbol of national power, national vitality in the 20th Century, and that the United States, if that was the symbolic quotient of national power, had to be first there.*

Mudd: *In the considerations that led up to the decision, was an alternative ever discussed about doing something other than going to the moon?*

Logsdon: *Yes, but only peripherally. Kennedy at one time during the decision-making process told his science advisor, Dr. Weisner from MIT, that if there were some other scientific spectacular which was as dramatic and convincing of American power, something like de-salinization, or nation-building, that maybe the United States should do that. But he came to the conclusion that space was the single most convincing demonstration of our technological power and that the decision then was what space program would allow us to beat the Russians in a spectacular fashion.*

As made plainly evident in Gerald Miller's earlier report on the reactions of young people in Rome, not everyone was enthusiastic about this day. Two young Americans with negative views, Gloria Steinem, a writer and contributing editor to *New York* magazine, and Ira Magaziner, a Brown University student activist who had succeeded in bringing about certain changes in the school's structure, were with Harry Reasoner in New York.

Reasoner: *Miss Steinem, is there something to you in this day's events that leaves your pleasure mixed?*

Miss Steinem: *It's hard to explain. I've been trying very hard to get enthusiastic about this day's events. Perhaps it's the kinds of stories I've been covering lately, which have been migrant workers and welfare recipients and those parts of our cities that looked as if they were remnants from World War II. The deserted buildings and everything. I was thinking about it this morning because it's as if we're getting more and more like 15th-Century Spain. We're discovering a new world but I wonder if we don't have our own Inquisition going in Vietnam in the name of that great religion of anticommunism. I'm sure we've napalmed many thousands more people than the Spanish ever burnt at the stake. And I wonder if a few centuries from now we won't be remembered for our time a little bit like 15th-Century Spain. I hope not.*

Reasoner: *Does that kind of thing express your opinions, Ira?*

Magaziner: *I don't really think so, completely. I used to be very, very excited about the space program. It was perhaps the most exciting thing to me when I was about ten years old. I even sent away a few cereal box tops and got assured that I would be on the first commercial flight to the moon. But I think over the past few years I can't really be as excited as I would like to be. I'm very enthusiastic about the program but there are just too many things that won't let me be really excited about it. For instance, while we're patting ourselves on the back about our great technology, and it is great, I can't help but think about the air pollution and water pollution which have resulted from that technology on earth. And it just disturbs me and I can't get very excited about all the money we spent and precautions that we've taken to protect the astronauts, which certainly is good and necessary but when we really don't even blink that much of an eye about a million people being killed in Biafra, it just makes me very ambivalent, I guess. I am very excited, certainly, today; but I just can't be as excited as I'd like to be.*

Reasoner: *You're saying somewhat the same things but also somewhat different. And to take yours up first, Miss Steinem, even assuming that most of what we're doing overseas is awful, isn't it still better to do something like this, which is not napalming anyone or injuring anyone and is advancing human knowledge. Isn't it a good thing to do?*

Miss Steinem: *Oh, yes, and I'm not saying that I would not do it if I were suddenly, magically, given the decision about priorities. It just seems that we could go a little more slowly. It seems that there's been a big change from 1961 when President Kennedy initiated this program and got it through Congress. In these eight years the Government has become a lot more estranged, I think, from the people, not just from the students and from the minorities but from the great white middle class and lower middle class who are overtaxed and overworked. And I think our priorities have to shift a bit.*

The lunar touchdown was now less than two hours away. The tension was building as the clock reached toward the apex of one of man's greatest adventures. There had been other great adventures—Hillary on Mt. Everest, Peary at the South Pole, the Wright Brothers at Kitty Hawk—but one of the most interesting was that of Sir Francis Chichester, who, alone, at the age of 65, had sailed his 53-foot boat, the *Gipsy Moth*, around the world in 266 days. Sir Francis was in CBS News' London studio with Mike Wallace, where he talked about the Apollo 11 astronauts and the moon landing.

Chichester: *They've got something I envy them enormously for. And that is that one of them is going to pull off one of the greatest firsts in history. I think that anybody who likes to go in for these kinds of adventures really values doing something that's not been done before. And here is the great historical first which we have coming up soon.*

Wallace: *If the opportunity were offered to you, would you have any desire to go into space yourself?*

Chichester: *I'd like to be the first man to step on the moon. I don't think* *many people in the world would refuse that privilege. But on the whole, I'm not very keen on it. For one thing, you're handing your initiative and enterprise to a ground crew, which I don't like. I like some project where you depend on your own initiative and control.*

Wallace: *Would you imagine—I know, of course, it's almost impossible and difficult for you to say—that Armstrong, Collins and Aldrin have a healthy fear?*

Chichester: *In their case, they're doing something very important at which Americans are so extremely good and that is they've drilled in every maneuver and in every act really efficiently. And when you practice a thing often you remove fear because you think it's going to be all right. Well instead, if something goes wrong or you think it's going to go wrong, you're frightened. And lastly, if it does go wrong, as I said before, you stop being frightened and you're intensely concentrating on how to put it right.*

Less than one and one-half hours before the scheduled time of touchdown, an updated report on the progress of Luna 15 came from Morley Safer and Sir Bernard Lovell at Jodrell Bank. Their information indicated that Luna 15 had been put into an orbit only ten miles above the moon's surface, and that the unmanned spacecraft was following an orbital path similar to that of Apollo 11. Sir Bernard had been surprised by the change of orbit: "It is very odd that the Russians should have changed the orbit of the probe so close to the time of the separation of the Apollo capsules. But, whether this has any significance, or is a coincidence, remains to be seen." He also noted that Luna 15 could not stay in lunar orbit very long at that height. The suspicion grew that the Russians were about to attempt to land Luna 15.

The Eagle's actual insertion into a descent orbit for the approach to the landing site was scheduled for 3:12 p.m. EDT. The firing of the descent engine would lower Armstrong and Aldrin to a 60-mile by 50,000-foot orbit, where they would remain until 4:08 p.m. when they would be given the "GO!" for a final approach to the landing site.

Within an hour man would be landing on the moon. One of the difficulties Armstrong and Aldrin would encounter in the final stages of the descent would be their passage through the sunbeam lighting the landing site. The sun's rays are so brilliant at certain points on the lunar surface that they literally wash out all topographical detail. George Herman, at the United States Geological Survey installation in Flagstaff, discussed the problem with astrogeologist Elliot Morris.

Herman: *As the Lunar Module begins the descent it will come in, in effect, through the particular sunbeam which is shining on the landing site. First it will be above that particular sunbeam, then as it comes down, it'll be right in the path of that sunbeam, then it will be below it. Dr. Morris, what is the problem as they pass through that particular beam?*

Morris: *Well, as they go through this point, the lunar surface whites out. There's no detail that can be seen at this point. We call that the Zero Phase point. As they pass beneath that sunbeam, the Zero Phase point then will move off from the landing site, and they'll be able to see the detail at the landing site.*

Herman: *So, in short, as they come down, as the Lunar Module slowly makes its descent towards the lunar surface, they will be able to see the lunar landing place, the landing site, in great detail. Then for just a few moments, a few critical moments, as they pass through the particular sunbeam that is illuminating the landing site, the site will be blanked out. And as they come down below the sunbeam, they will be able to see the site once again in even sharper detail. That's the nature of the problem.*

One hour was left until Eagle touched down on the moon. Cronkite and Schirra noted that the "LM is cramped in comparison to the Command Module; there's just enough room for Neil and Buzz." To Cronkite's comment that while it was "not exactly a homey place to stake out," Schirra noted that the astronauts had an important four-hour rest period prior to the walk on the moon. Cronkite commented that "none of us slept well. Even the astronauts slept four or five hours instead of the scheduled eight on the flight plan." Cronkite by this time was pretty well convinced that there would be an early walk.

Fifty minutes to touchdown. Heywood Hale Broun was with some interested spectators at Disneyland, at the site of a moon ride.

Broun: *There is more than one kind of flight to the moon. And this one is one which 1600 people an hour can take. It gives you the idea of what it would be like to go, if, perhaps, you are not actually going. Members of the Russian track team, who have been here in a two-day meet, are with me now. I'll ask one of them—Mr. Terevanesian, with both the Soviet ship and the American ship out there, is this rather like a track meet?*

Terevanesian: *It seems to me that this is like a competition. But there will be no losers. Both sides will win.*

3:31 p.m. EDT, and it was just over 45 minutes to touchdown. Cronkite and Sevareid sat at the anchor desk talking about the significance of the moment.

Cronkite: *Eric, we're getting close to that time now. It's just 45½ minutes from the touchdown on the moon's surface. One of the trite questions of young interviewers since time immemorial is "How do you feel?" But I say it anyway. How do you feel? I'm getting knots in my stomach.*

Sevareid: *Well, I haven't got to that point yet. I don't think the human mind can stay tense, with its mind directly on focus, terribly long. I don't know how long the average human attention span is.*
 The mind wanders in search of all kinds of strange odds and ends. I find myself thinking about this Lunar Module and why it looks like an insect. It looks terribly non-functional, and of course it's nothing but functional. Not functional in more pedantic terms, like earth; but 71

in its own terms—and maybe it has to look like an insect. Some ento-
mologists, you know, think insects will take over the earth, and do
away with man one day. So maybe it's natural this thing does look like
a bug. All these paradoxes come to your mind as you read and you brood
about it a bit. Here we are worrying about germs and microbes from
the moon coming to pollute the earth, when we and no doubt the other
space power, the Soviet Union, are building up vast stores of bacteri-
ological warfare weapons—germs, that can destroy millions of people.
This seems rather incredible. But certainly, it's highly more likely to
pollute the moon, if that matters.

Also we talk about earth resources—or moon resources—because
we don't know half of what's in the earth now. And as somebody
pointed out, if the moon's surface were covered with polished diamonds,
it wouldn't be economic to go and retrieve them. It would just cost too
much. Well, I'm sure we're going to add it all up one day. But philosophy
will have to come later. I want to see it work first, and I'm sure you're
in the same state of mind.

Cronkite: *Well, we're going to be able to get a chance to see it work, just*
in this remarkable day, and this remarkable age, in 41 minutes from
now. You know, I'd just like to make the point once more, because so
much was made of it last week down at the Cape. One of the reasons
that Dr. Charles Berry of the Space Program, didn't want extra people
coming into his little laboratory, where he had the spacemen pretty well
sealed up, that last week before they left, was not because he thought
they'd give them anything necessarily, but he wanted to keep a monitor
on what they themselves were taking to the moon. So he'd know not
only what was left up there in the way of diseases, but what they had
acquired and brought back. It's hard to control this thing, but actually
that contamination of the moon's surface is serious.

The tension increased during the last 40, then 30 minutes before
touchdown. Cronkite and Schirra had been joined by Newell Trask,
the United States Geological Survey astrogeologist who had coordinated the mapping of landing site #2. Trask was asked what he

thought Armstrong and Aldrin would find when they stepped onto the lunar surface.

Cronkite: *Are they going to sink in moon dust, or are they going to, with their one-sixth gravity, sort of bound across it?*

Trask: *They'll move very slowly, naturally, for the sake of caution. Probably over all of the moon is a fine-grained, soil-like material. This will support their weight, so there's no danger of them sinking into it.*

Cronkite: *They'll sink a little bit. It's spongy, is it not?*

Trask: *Yes, they'll sink a little bit. Wet beach sand has been used repeatedly as an analogy, and I guess that's right.*

Cronkite: *What about this business we've heard of the moon dust clinging to them? Perhaps obscuring their vision by clinging to their visors. What is that problem?*

Trask: *Well, it doesn't appear to do that. It didn't do it on the Surveyor spacecraft to any great extent. When dust was kicked up by the firing of the engine, then it collected on some parts of the Surveyor. But Armstrong and Aldrin won't get out, of course, until the engine is completely shut down.*

Eagle was on the back side of the moon. When it came around again, and Houston acquired the spacecraft's signal, Armstrong and Aldrin would be at an altitude of approximately 10 miles, on a downward sweep to the moon's surface. The word everyone had been waiting for came from Houston.

Capcom: **Eagle, Houston. If you read, you're a go for powered descent. Over.**

Columbia: **Eagle, this is Columbia. They just gave you a go for powered descent.**

Capcom: **Columbia, Houston. We've lost them on the high gain** antennae again. Would you please—we recommend they yaw right 10 degrees and try the high gain again.

Columbia: Eagle, you read Columbia?

Eagle: Roger, read you.

In Eagle, Buzz Aldrin was reading through a checklist with Neil Armstrong.

Schirra: *It's a double confirmation to insure that everything is going well.*

Cronkite: *Two and a half minutes now to ignition…14 minutes to touchdown…engine fires for 12 minutes.*

Houston: Coming up one minute to ignition.

Cronkite: *One minute to ignition, and 13 minutes to landing. I don't know whether we could take the tension, if they decided to go around again. 40 seconds. 30 seconds.*

Houston: Current altitude about 46,000 feet, continuing to descend.

An animation of the power descent firing sequence flashed on the screen.

Capcom: Eagle, Houston. Everything is looking good here. Over.

Cronkite: *I think you can hear the excitement in his voice.*

Schirra: *We're in real good shape again.*

Cronkite: *Ten minutes to the touchdown. Oh, boy! Ten minutes to a landing on the moon.*

Schirra: *To think it's been almost 10 years since we've been trying to do this.*

Capcom: Eagle, Houston. You are go. Take it all at four minutes. Roger you are go—you are go to continue power descent. You are to continue power descent. Altitude 40,000 feet. We've got data dropout. You're still looking good.

Eagle: And the earth right out our front window.

Cronkite: *Got about 68 miles to go. Seven minutes left before landing.*

Houston: We're still go. Altitude 27,000 feet.

Eagle: Throttles down better than the simulator.

Capcom: Rog.

Schirra: *He says it's better than the simulator. That's consoling.*

Cronkite: *Crowds around this country and all over the world are watching this and listening to these communications.*

Houston: 21,000 feet, still looking very good.

Cronkite: *They've got just 14 miles to go and 4½ minutes. Four and a half minutes left in this era.*

Capcom: Altitude 13,500. Eagle, you're looking great at eight minutes. Correction on that velocity, now reading 760 feet per second.

Schirra: *760 feet per second. They're on the way down. That's pretty slow for space flying.*

Cronkite: *That is as slow as man has ever flown in space.*

Schirra: *It sure is.*

Capcom: We're go. Altitude 9200 feet. You're looking great. Descent rate 129 feet per second.

Cronkite: *7600-foot level. 1.4 miles high, when they get down to a speed of 98 miles an hour. And they're just a little under five miles from the landing site. At that point they can pitch forward.*

Capcom: Eagle, you're looking great, coming up on nine minutes. We're now in the approach phase, everything looking good. Altitude 5200 feet.

Cronkite: *5200 feet. Less than a mile from the moon's surface.*

Eagle: Manual altitude control is good.

Capcom: Roger. We copy. Altitude 4200 and you're go for landing. Over.

Eagle: Roger, understand. Go for landing. 3000 feet. Second alarm.

Cronkite: *3000 feet. Um-hmmm.*

Eagle: Roger. 1201 alarm. We're go. Hang tight. We're go. 2000 feet. 2000 feet, into the AGS. 47 degrees.

Cronkite: *These are space communications, simply for readout purposes.*

Capcom: Eagle looking great. You're go.

Houston: Altitude 1600. 1400 feet. Still looking very good.

Cronkite: *They've got a good look at their site now. This is their time. They're going to make a decision.*

Eagle: 35 degrees. 35 degrees. 750, coming down at 23. 700 feet, 21 down. 33 degrees.

Schirra: *Oh, the data is coming in beautifully.*

Eagle: 600 feet, down at 19. 540 feet down at 30—down at 15...400 feet down at 9...8 forward...350 feet down at 4...300 feet, down 3½...47

10:56:20
A Pictorial Essay

The Launching July 16

6:08
Schirra and Cronkite at Cape Kennedy

6:28

6:05 a.m.

Breakfast (Recorded)

6:37

MAN
ON THE
MOON

6:06

6:27

LAUNCH PAD 39A

6:48

THE EPIC
JOURNEY OF
APOLLO 11

6:07

6:49

8:59
Sevareid and Cronkite

9:04
Heywood Hale Broun on the beach

9:31:58

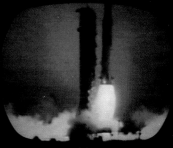

9:32

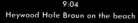

9:10

9:31:59

9:32

VOICE OF JACK KING
LAUNCH CONTROL

9:31
30 seconds and counting

9:32
"Lift-off"

9:32

9:32
Johnson and Agnew

9:32
"Tower cleared"

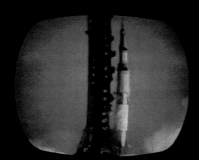

9:32
"We have a lift-off."

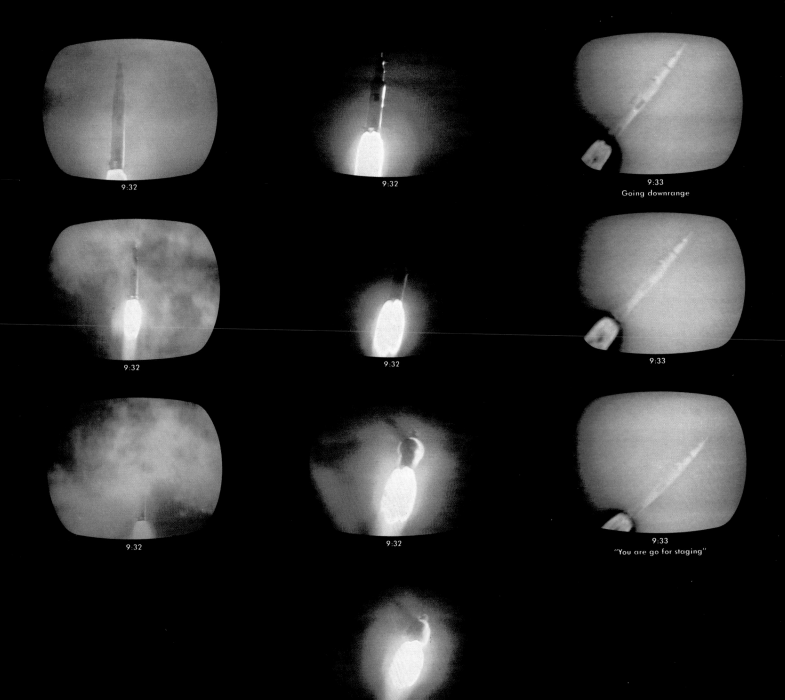

9:32

9:32

9:33
Going downrange

9:32

9:32

9:33

9:32

9:32

9:33
"You are go for staging"

9:32

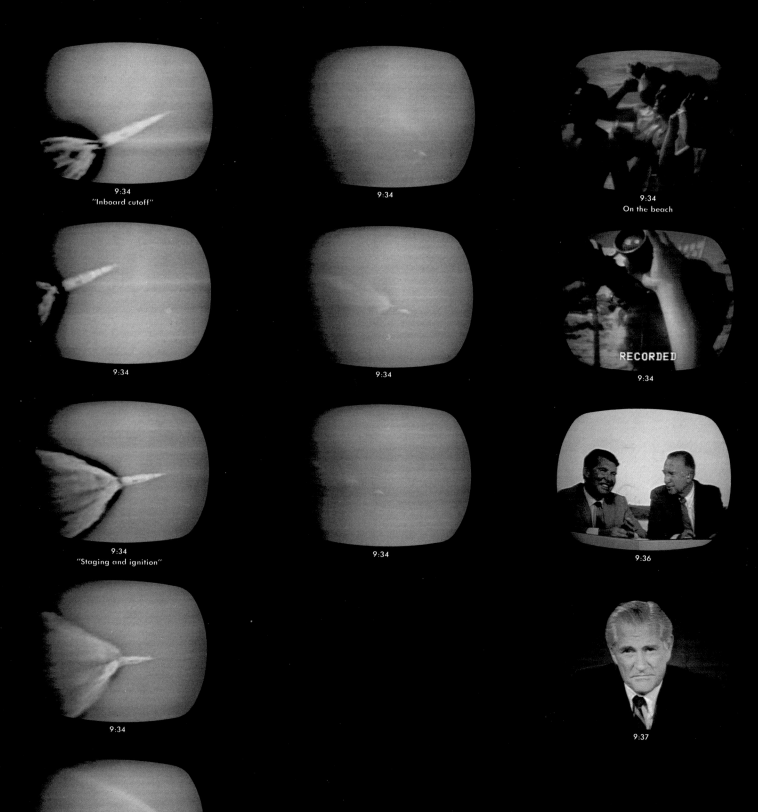

9:34
"Inboard cutoff"

9:34

9:34
On the beach

9:34

9:34

9:34
RECORDED

9:34
"Staging and ignition"

9:34

9:36

9:34

9:37

9:34

9:39
Second stage separates, drops away

9:40
Third stage ignites

9:40

9:41

10:02
Vice President Agnew

10:06

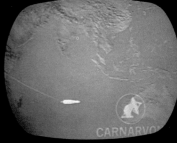

10:13

10:19
Former President Johnson

10:26

10:37
...at Grumman

NELSON BENTON
CBS NEWS

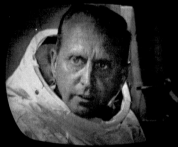

10:39
Scott MacLeod at Grumman

10:44
Leo Krupp at North American

10:47
Dr. Krafft Ehricke and Stout

11:20
The moon walk simulated at Grumman

11:21

11:21

11:23

11:26

11:27

11:28

11:28

11:40
Arthur C. Clarke and Cronkite

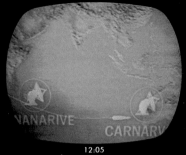
12:05

11:29

11:45

12:06
Dr. Ralph Abernathy at lift-off (tape)

11:30

12:03 p.m.
Bruce Morton at Houston

12:07

11:31
Replica of plaque to be left on the moon

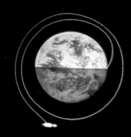

12:20

12:49

CBS NEWS SIMULATION

12:53

12:24

12:50
Command Module separates...

12:57

12:25
Cronkite reminisces about old days at the Cape

CBS NEWS SIMULATION

12:52

CBS NEWS SIMULATION

12:59
...docks with Lunar Module
and pulls away from third stage

12:33
Schirra describes lapel pin

CBS NEWS SIMULATION

12:52

12:53

On the Way

Live Transmission July 17

7:46
Michael Collins

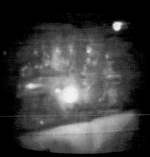

7:56
The astronauts demonstrate zero gravity...

7:33 p.m.

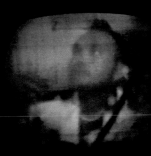

7:47

7:57
...by passing a flashlight back and forth

7:34

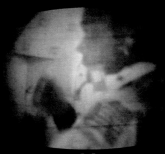

7:47

7:58

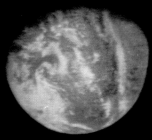

7:39

7:58

7:59

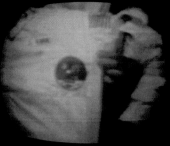

8:00
The Apollo 11 patch

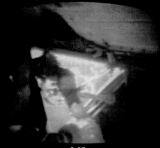

5:57
Aldrin attaches camera in LM window

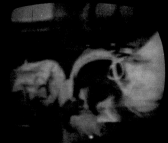

6:04
Aldrin describes his helmet

Live Transmission July 18

5:58

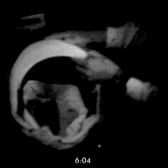

6:04

5:55 p.m.
Aldrin in Eagle

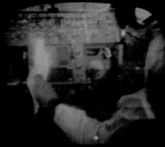

6:00
Control panel

6:04

5:56

6:02
Control panel

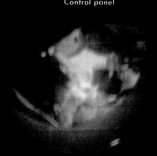

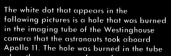

The white dot that appears in the
following pictures is a hole that was burned
in the imaging tube of the Westinghouse
camera that the astronauts took aboard
Apollo 11. The hole was burned in the tube

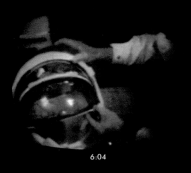

6:04

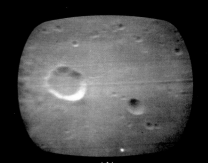

4:14

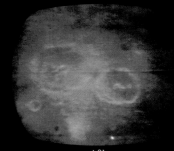

4:01

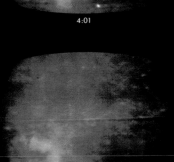

6:04

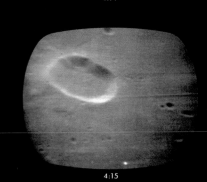

4:15

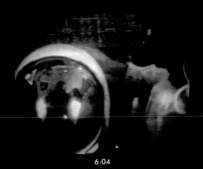

6:04

Live Transmission July 19

The astronauts gave the world a view of the moon and their landing site from a 100-mile-high orbit as they approached the 24-hour mark to their landing.

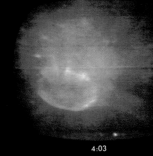

4:02

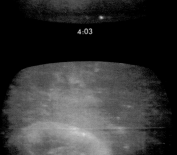

4:17

240,000 MILES FROM EARTH

3:51 p.m.
The astronauts in lunar orbit

4:03

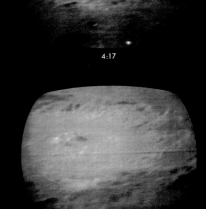

4:18

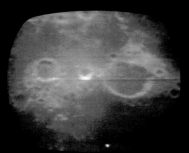

4:10

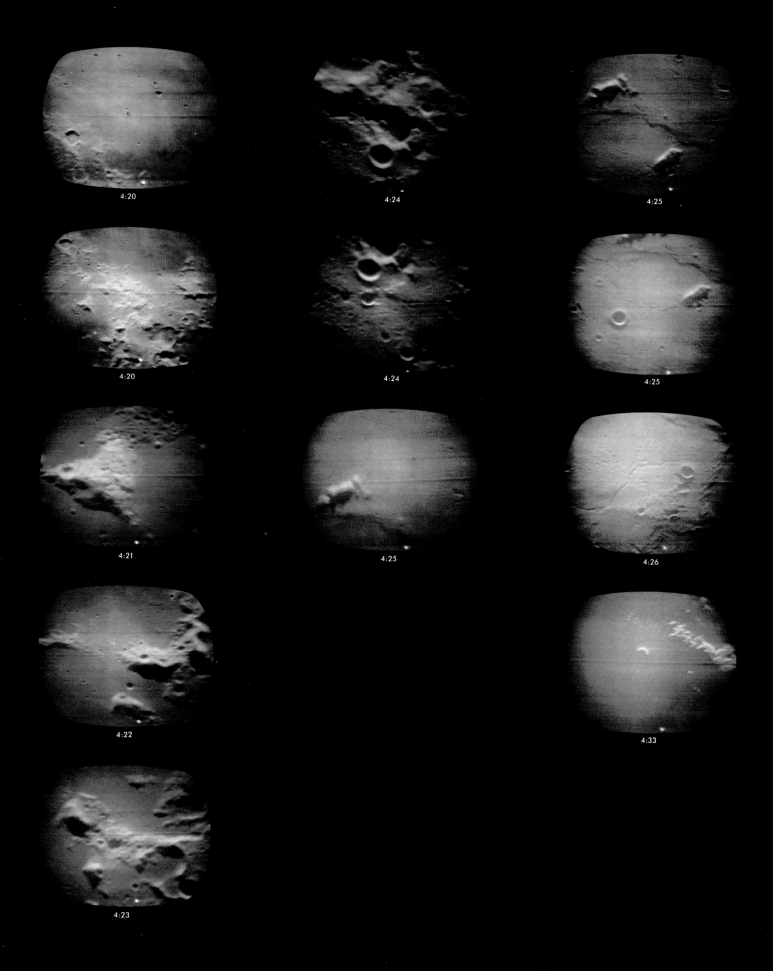

Man on the Moon July 20, 21

Using an animated film, Charles Kuralt opened the historic broadcast with a poetic description of the creation of the moon and its effect on life on earth.

11:03

11:04

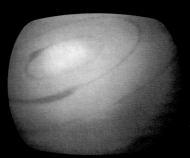

11:00 a.m.

11:04

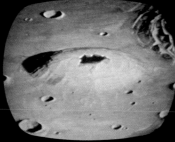

11:05

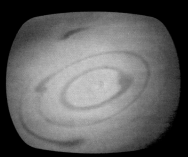

11:02

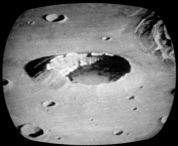

11:04

11:05

11:02

11:06

11:02

TIME TO LUNAR LANDING

11:07
Cronkite in place in New York

11:21
George Herman at Flagstaff

TIME TO LUNAR LANDING
4 : 4 7 : 5 3
HRS   MIN   SEC

11:29

11:15
Bruce Morton in Houston

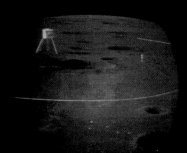

11:22
John Hart at Cinder Lake

11:30

11:17
Marya McLaughlin in Houston

VAN CORTLANDT PARK

11:23
Bill Plante in New York

11:30

11:30

11:45
Dr. Goldmark explains the CBS color...

11:54
At the White House

11:58
...with Morley Safer at Jodrell Bank

11:47
...field sequential television system...

11:57

12:02 p.m.
George Syvertsen in Tokyo

11:47
...used aboard Apollo 11

12:15
Astronaut Tom Stafford with Bruce Morton

11:51
Newell Trask, Arthur C. Clarke,
Schirra and Harry Reasoner

12:41
Bill Haskell at Mystic Seaport, Conn.

1:30
Mike Wallace

1:46
The undocking

12:58
Don Wayne in Dayton

1:32
...in Paris

1:47
"The Eagle has wings."

1:03
Barry Serafin in St. Louis

1:36
Marvin Kalb in Bucharest

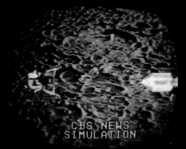

1:57

1:17
Kurt Vonnegut, Jr. joins Clarke and Reasoner

1:38
Gerald Miller at Rome's Spanish Steps

2:41
Ira Magaziner and Gloria Steinem dissent

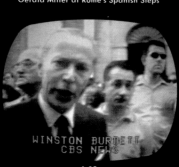

1:39
...reporting from Rome

'R FRANCIS CHICHESTE'

2:48
with Mike Wallace in London

48:45

MIN   SEC

3:17

3:39
Astrogeologist Newell Trask

ANIMATION

3:12
Armstrong and Aldrin
begin their descent to the moon

DR. ELLIOT MORRIS
ASTROGEOLOGIST

3:20
...with George Herman at Flagstaff

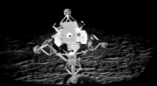

3:48

3:31

TIME TO LUNAR LANDING

23:00

MIN.   SEC.

3:53

3:34

3:34

4:00
Mission Control

4:05

4:10

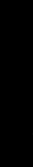

4:02
"Go for powered descent"

4:05

POWER DESCENT

ALTITUDE   28,000 FT

VELOCITY   1,800 FPS

RANGE TO GO   34

4:11

TIME TO LUNAR DESCENT

14:55

MIN.   SEC.

4:03

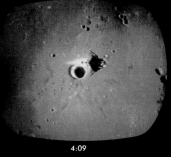

4:06

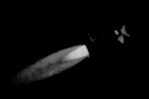

4:12

4:03

4:09

TIME TO LUNAR LANDING

00:17

MIN.   SEC.

4:16

TIME TO LUNAR LANDING

07:00

4:16

4:16

4:16

4:16

4:17:42
"Man on the moon!"

4:18

4:18

4:18
"I've been saying them all under my breath."

4:19

4:19

4:19
"Whew, boy—"

4:19

LUNAR MODULE HAS
LANDED ON MOON

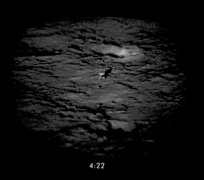

4:22

4:57
Mr. and Mrs. Stephen Armstrong

5:10

LIVE VOICE OF
ASTRONAUT ALDRIN

4:27

4:57

JFK AIRPORT

5:12
Joseph Benti at Kennedy Airport

4:29

4:57

DISNEYLAND

5:18
Broun at Disneyland

4:57

DR. TERRY OFFIELD
ASTROGEOLOGIST

5:42
...with George Herman at Flagstaff

THOMAS PAINE

5:50
...at Mission Control

6:10
Mrs. Neil Armstrong

6:26
Prime Minister Harold Wilson

SMITHSONIAN
INSTITUTION

6:44
Astronaut Frank Borman joins Rather and Mudd

6:11

LIVE VIA SATELLITE

6:30
Daniel Schorr in Amsterdam

NEW YORK CITY

6:58
Plante at Harlem soul festival

6:14

TRAFALGAR SQUARE

6:31
London

ATLANTA, GEORGIA

7:08
Ray Moore in Atlanta

PARIS, FRANCE

6:32
Paris

7:18
Mexico City

BELGRADE
YUGOSLAVIA

6:37
William McLaughlin in Belgrade

7:23
Mrs. Michael Collins

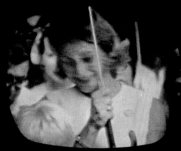

7:25
Mrs. Buzz Aldrin

8:15
Roger Mudd with Vice President Agnew

7:24

7:25

8:24

7:24

7:25

ROBERT HEINLEIN

8:29
...at North American Rockwell

7:25

10:11

10:50

10:54

10:11

10:54

10:54

MISSION CONTROL

10:49

10:54

10:55

10:50

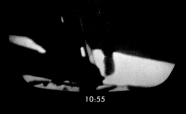

10:55

10:55

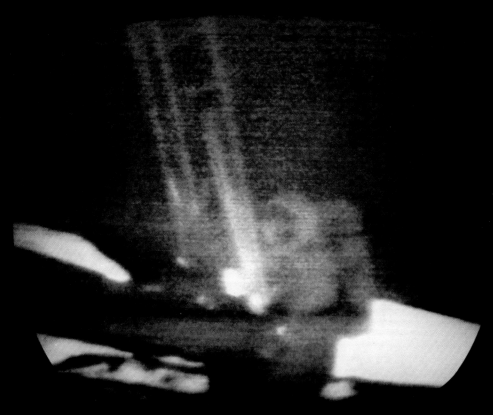

10:55

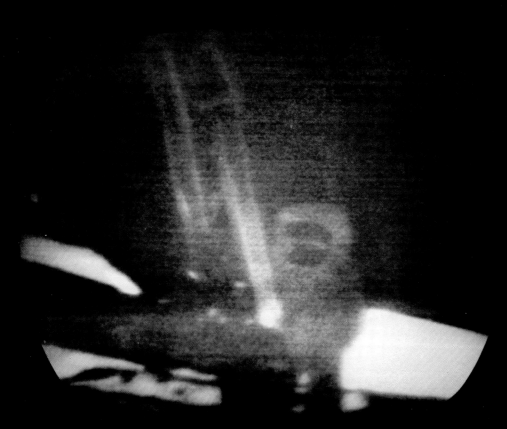

10:56:20

"That's one small step for a man..."

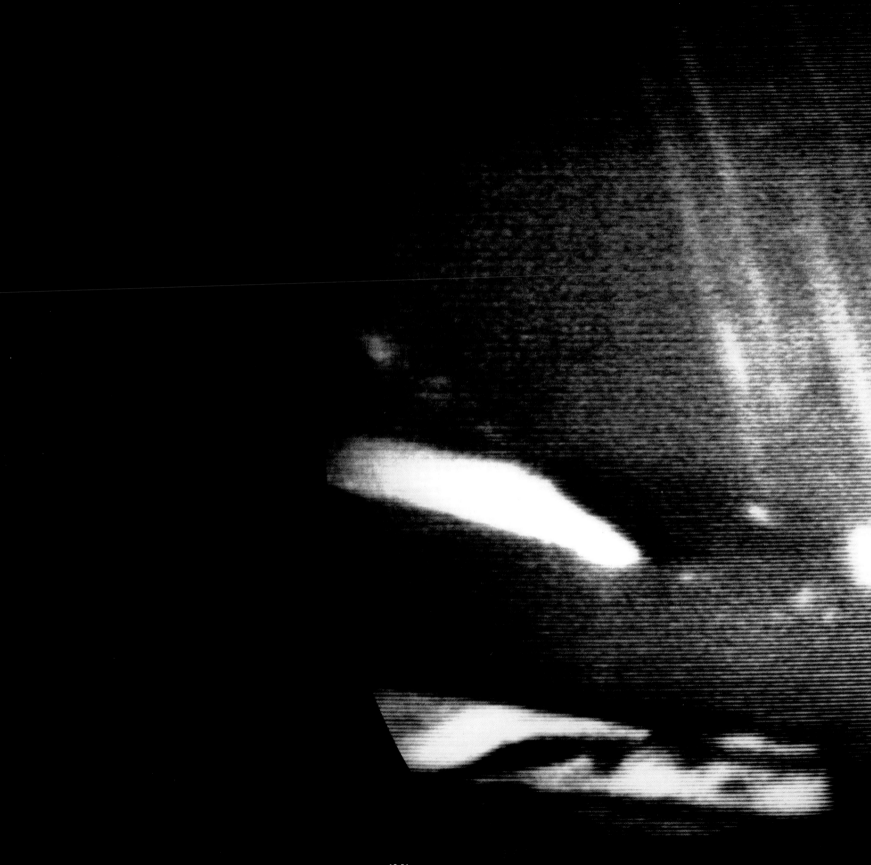

10:56
"...one giant leap for mankind."

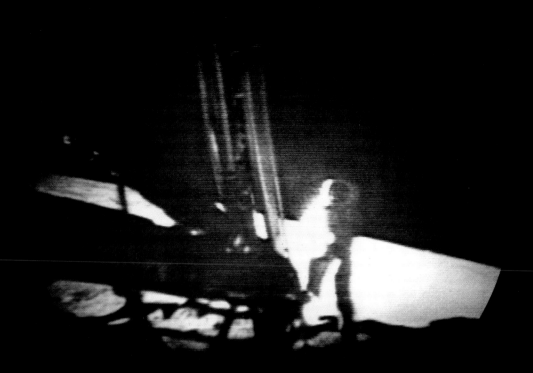

11:11
Armstrong guides Aldrin down the ladder

11:14
Aldrin descends

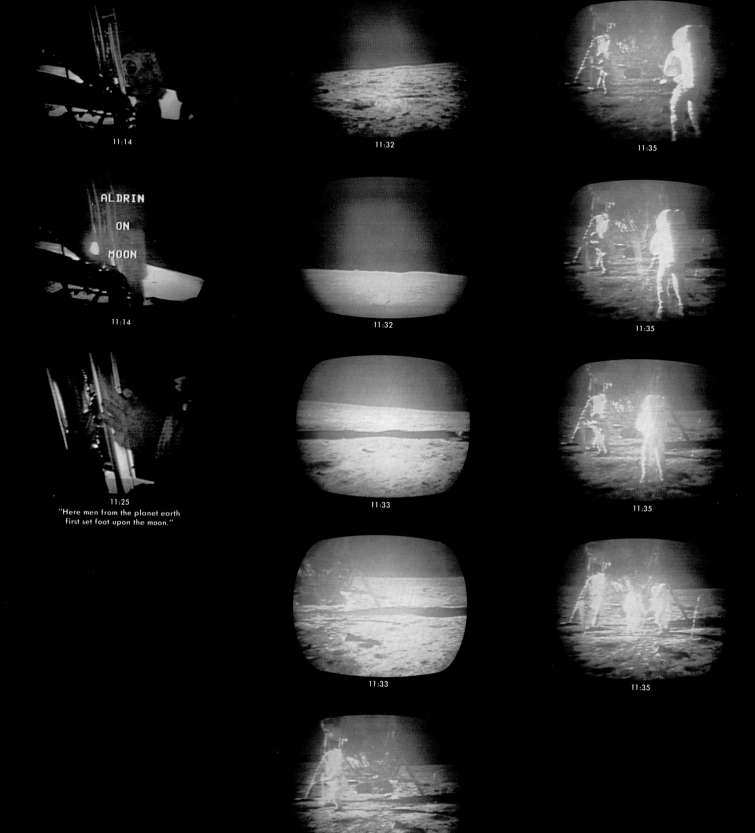

11:14

11:14

ALDRIN
ON
MOON

11:25
"Here men from the planet earth
first set foot upon the moon."

11:32

11:32

11:33

11:33

11:34

11:35

11:35

11:35

11:35

11:36
The flag is unfurled

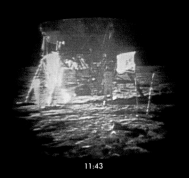

11:43

11:46

11:39

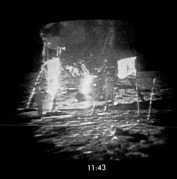

11:43

11:46

11:40

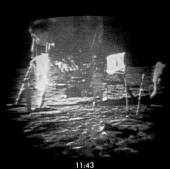

11:43

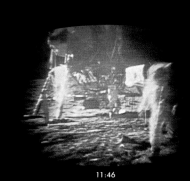

11:46

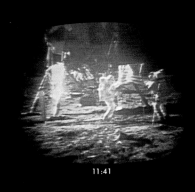

11:41

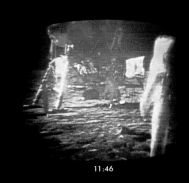

11:46

11:46

11:47

12:48 a.m.

2:02

11:48
President Nixon talks with the astronauts

12:50

2:07

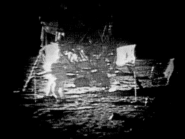

LIVE

1:10
Monday, July 21...the walk is over

WILLIAM A. ANDERS

2:17

1:34
Mrs. Stephen Armstrong

6:19 a.m.
...with Frank Kearns

8:16
Winston Burdett in Rome

8:20
David Schoumacher in New York

6:22
Stonehenge

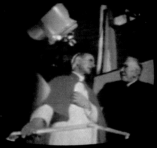

8:17
Pope Paul VI

8:21
Biography of Neil Armstrong

6:36
Da Vinci

8:17

8:21

6:44
Apollo 1 crew members Virgil "Gus" Grissom,
Edward White and Roger Chaffee
were killed in the tragic
Apollo 204 fire at Cape Kennedy on
January 27, 1967. A memorial to
the crew was left on the lunar surface.

8:18

8:18

9:13

9:25

Mudd with Senator Charles Percy

11:35

9:13

10:22

Richard C. Hottelet with Arthur Goldberg

11:37

9:13

10:23

11:43

9:13

11:04

Filmed interview with former President
Johnson at the LBJ ranch

11:12

11:49
Mike Wallace and Ray Bradbury

12:55

1:54

RAY BRADBURY

11:53

SOUTH VIET-NAM

12:59
Soldier in South Vietnam,
Don Webster reporting

1:54

12:20 p.m.
Harry Reasoner with Vonnegut,
Dr. Menninger, Rev. Gill and Commager

1:54

1:54
Lift-off

1:55

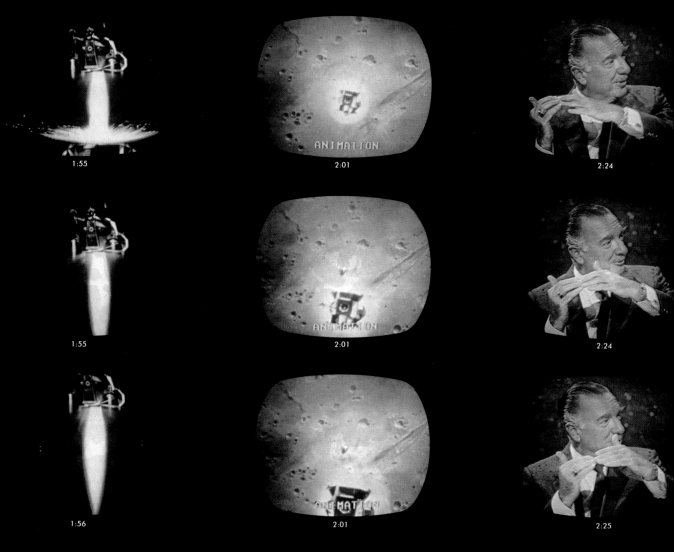

1:55

1:55

1:56

2:01

2:01

2:01

2:24

2:24

2:25

LUNAR MODULE

LM-CSM RANGE

290 ST.MI.

2:04

3:09

4:22

Harry Reasoner on moon lore

Science fiction movie sequence...

DR. THOMAS GOLD
NEW YORK

3:15

4:25

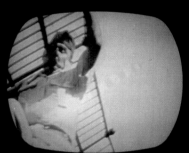

...narrated by Orson Welles

DR. HAROLD UREY
HOUSTON

3:15

4:25

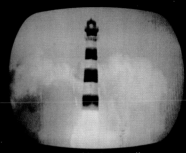

U.S. Air Force film   1960

DR. GERARD KUIPER
FLAGSTAFF

3:15

4:25

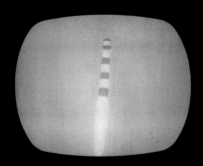

LUNAR MODULE

COMMAND MODULE

4:35

"A Trip to the Moon" — 1902

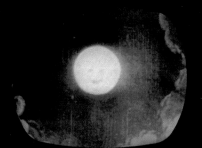

"Woman on the Moon" — 1929

'10 seconds to go!'

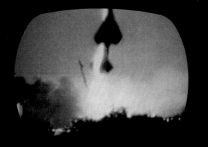

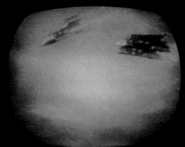

"Flash Gordon" — 1938

5:24
The rendezvous

"Destination Moon" — 1950

5:25

"Barbarella" — 1968

CBS NEWS
SIMULATION

5:25

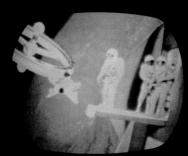

"The Conquest of Space" — 1955

5:18
Clarke equates the reality of the past five
days with the fantasy of science fiction

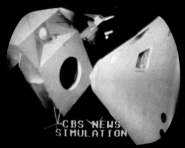

5:32

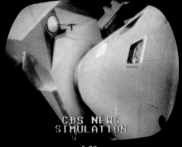

5:32

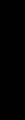

5:35
Eagle and Columbia are one

5:43

5:43

5:45

LEO KRUPP

5:45

5:46
Herman

5:50
MacLeod and Benton

Homeward Bound
Live Transmission July 22

It was 12:56 AM, and Neil Armstrong had fired the Command Service Module engine to propel the spacecraft out of lunar orbit. Apollo 11 was on its way back to earth.

Later that day the astronauts gave earthlings a lesson in how to operate in the Zero-G environment of space. Buzz Aldrin made a ham spread sandwich and then applied the principle of the gyroscope with the empty tin.

9:17
Aldrin and a piece of bread

9:20
The gyroscope

9:18

9:20

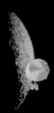

LIVE
VOICES OF ASTRONAUTS

12:56 a.m.

9:18

9:20

9:02 p.m.
Mission Control

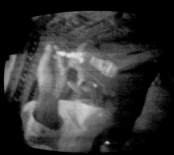

9:20

9:22
Collins demonstrates, with a spoonful of water, the effect of weightlessness.

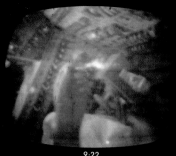

9:22

7:12
Buzz Aldrin

7:14

Live
Transmission
July 23

7:12

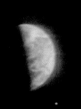

7:17

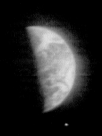

7:04 p.m.

7:14
Neil Armstrong

7:18

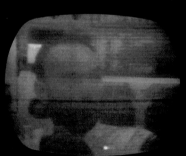

7:08
Mike Collins

7:14

Splashdown
July 24

TIME TO SPLASHDOWN
4 9:4 5
MIN.   SEC.

12:02

ANIMATION

12:25
Re-entry

12:00 noon

DALLAS TOWNSEND
CBS NEWS ABOARD
THE USS HORNET

12:04

12:25

12:01

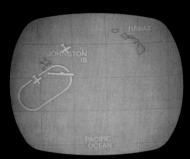

12:16

TIME TO SPLASHDOWN
1 3:1 4
MIN.   SEC.

12:37

12:02

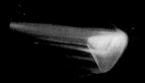

12:39
Columbia sighted

12:39

RON NESSEN
ABOARD THE USS HORNET

TIME TO SPLASHDOWN
0 6:06

12:44

12:53
President Nixon

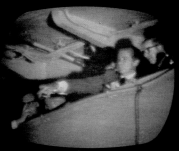

12:55

TIME TO SPLASHDOWN
0 3:04

MIN.   SEC
12:47

LIVE FROM
THE USS HORNET

12:53

12:56

TIME TO SPLASHDOWN
0 1:52

MIN.   SEC
12:48

12:53

COLUMBIA

1:10
Columbia

COLUMBIA

HAS

RETURNED

12:50:35
Splashdown

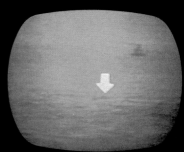

1:11
Columbia

1:45

1:47
Frank Borman

1:50

1:46

1:50

1:50

1:47

1:50

1:51

1:47

1:51

1:52

1:54

1:57

2:37

1:56

1:58
Mission Control

2:38

1:57

2:00
Helicopter landing on ship

2:39
The Mobile Quarantine Facility

2:01
Helicopter and crew

2:49

2:55

2:56
Aldrin

2:53
The President...

2:55

2:56
Armstrong

2:54
...on his way to greet the astronauts

NEIL ARMSTRONG
SPACECRAFT COMMANDER

2:55

2:56
Collins

EDWIN ALDRIN
LM PILOT

2:55

2:56

3:04

3:29

3:02

3:04

3:32

3:02
Collins

3:32

3:02
Armstrong

3:32

3:03
Aldrin

forward…1½ down…70…got the shadow out there…50, down at 2½, 19 forward…altitude-velocity lights…3½ down…220 feet…13 forward …11 forward, coming down nicely…200 feet, 4½ down…5½ down… 160, 6½ down…5½ down, 9 forward…5 percent…quantity light 75 feet. Things still looking good, down a half…6 forward…lights on…down 2½…forward…40 feet, down 2½, kicking up some dust…30 feet, 2½ down…faint shadow…4 forward…4 forward, drifting to the right a little…6…drifting right…

Cronkite: *Boy, what a day.*

Capcom: 30 seconds.

Eagle: Contact light. O.K. engine stopped…descent engine command override off…

Schirra: *We're home!*

Cronkite: *Man on the moon!*

Eagle: Houston, Tranquility Base here. The Eagle has landed!

Capcom: Roger, Tranquility. We copy you on the ground. You've got a bunch of guys about to turn blue. We're breathing again. Thanks a lot.

Tranquility: Thank you.

Cronkite: *Oh, boy!*

Capcom: You're looking good here.

Cronkite: *Whew! Boy!*

Schirra: *I've been saying them all under my breath. That is really something. I'd love to be aboard.*

Cronkite: *I know. We've been wondering what Neil Armstrong and Aldrin would say when they set foot on the moon, which comes a little bit later now. Just to hear them do it. Absolutely with dry mouths.*

Capcom: Roger, Eagle. And you're stay for T-1. Over. You're stay for T-1...

Tranquility: Roger. We're stay for T-1.

Capcom: Roger. And we see you getting the ox.

Cronkite: *That's a great simulation that we see here.*

Schirra: *That little fly-speck is supposed to be the LM.*

Cronkite: *They must be in perfect condition...upright, and there's no complaint about their position.*

Schirra: *Just a little dust.*

Cronkite: *Boy! There they sit on the moon! Just exactly nominal wasn't it...on green with the flight plan, all the way down. Man finally is standing on the surface of the moon. My golly!*

Capcom: Roger, we read you Columbia. He has landed. Tranquility Base. Eagle is at Tranquility. Over.

The words "Houston, Tranquility Base here. The Eagle has landed," brought a moment of tremendous emotional release. Cronkite sat speechless, glasses in hand, shaking his head from side to side. Schirra wiped a tear from his eye. Shouts and applause broke the stillness that had settled over the CBS News control room in the seconds before touchdown, and a cheer went up from the technicians working on the studio floor. There was so much noise on the floor that soundman Sam Laine had to ask Banow to ask for quiet. Cronkite had heard the technicians' cheer, and said later that he couldn't remember any other time when there was a similar outburst. It was an extraordinary moment for everyone, and in the days after would be mentioned time and again as the most memorable of the eight days.

For anyone who had been involved in space for any period of time, the landing would have the greatest impact. As dramatic and visually exciting as the walk would be, the landing was by far the most dangerous part of the mission. Joan Richman and Wussler both mentioned the fact that the landing made the greatest demands on the imagination. With only the voices of the astronauts as a guide, one had to visualize Armstrong at the controls in the final seconds, making the critical decision whether or not to set Eagle down on the lunar surface.

With Eagle down, Wussler made the decision to listen to communications between the astronauts and Houston. For the next 20 minutes, the visual elements of the broadcast would originate at Grumman, with the model of the Lunar Module sitting on the mock moonscape. Producer Frank Manitzas and director Al Mifelow were to say later that the request to "hold that one shot for 20 minutes" led to the longest 20 minutes they had experienced in television.

Schirra: *I'll bet they really had a ball when they saw that blue contact light come on.*

Cronkite: *That's the one that comes on when they're five feet eight inches from the surface. The probe tells them they're at five feet eight inches. And they turn the engine off. One of the first things they'll do when they climb down and start taking a look at the Lunar Module from the moon's surface is to look at those landing pods to be sure that all four of them have good contact. So we've got an unofficial time, which probably will end up being the official time, of setting down at 4:17:42. Seventeen minutes and 42 seconds after four o'clock, Eastern Daylight Time, on this day, July 20th, 1969.*

Schirra: *We have half of that major goal done. That's getting men on the moon by 1970. The other half is to get them off and bring them home. And this word "nominal" sounds so good to me, I have no trepidations whatsoever about the return.*

Cronkite: *The way it's gone, they certainly have built our confidence in these machines.*

Following a quick check-through of the LM's systems to assure that it had survived the landing intact, the crew began to describe the landing and the lunar surface.

Tranquility: Houston, that may have seemed like a very long final phase. The auto targeting was taking us right into a football field-sized crater, with a large number of big boulders and rocks for about one or two crater diameters around us, and it required flying manually over the rock field to find a reasonably good area.

Capcom: Roger, we copy. It was beautiful from here. Tranquility, over.

Tranquility: We'll get to the details of what's around here, but it looks like a collection of just about every variety of shapes, angularities, granularities, every variety of rock you could find. There doesn't appear to be too much of a general color at all. However it looks as though some of the rocks and boulders, of which there are quite a few in the near area, are going to have some interesting colors to them. Over.

Capcom: Rog, Tranquility. Be advised there're lots of smiling faces in this room, and all over the world.

Tranquility: There are two of them up here.

Capcom: Rog, that was a beautiful job, you guys.

Columbia: And don't forget one in the Command Module.

Then Mike Collins in Columbia had the chance to talk to Armstrong and Aldrin for the first time since the landing.

Columbia: Tranquility Base. It sure sounded great from up here. You guys did a fantastic job.

Tranquility: Thank you. Just keep that orbiting base ready for us up there now.

Eagle had been given the "GO" for a stay on the lunar surface. If Armstrong and Aldrin kept to the flight plan, Armstrong would become the first man to step on the lunar surface in just over nine hours, shortly after 2:00 a.m. EDT, on Monday, July 21.

Out in Wapakoneta, Ohio, Mr. and Mrs. Stephen Armstrong, the parents of astronaut Neil Armstrong, had stepped out onto the front lawn of their home. Pool reporter Mark Landsman interviewed them.

Landsman: *Mrs. Armstrong, what were your feelings at the moment that Apollo 11 was coming toward the moon?*

Mrs. Armstrong: *When it was coming toward the moon, well, I was just hoping and praying everything would go well. And at touchdown, I was saying "Praise the Lord from whom all blessings flow."*

Landsman: *Mr. Armstrong?*

Mr. Armstrong: *Very, very thankful that there was a successful landing.*

Correspondents Joseph Benti at the International Arrivals Building at Kennedy Airport in New York City, Heywood Hale Broun at Disneyland and Roger Mudd at the Smithsonian Institution were with large groups of people who had watched the landing on enlarged Eidophor television screens.

Benti: *When the news came here, about three minutes before it came, this building fell into a hush, and watched our large television screen up there. There were about two or three thousand people in here, just jam-packed. Eric Sevareid, I recall, said something about our inability to really concentrate over long periods of time. And that was true here. You never got the sense that anyone was waiting for the lunar landing until those three minutes just before touchdown. And then the feeling took a good hold of everybody, and they fell into a hush and we were only wishing that you'd get a shot of this, because they broke into applause and cheers, as the landing was announced. Actually, Walter, you begin getting the same phrases over and over again. I think everybody here would say it the same way, it was "fabulous" or "spectacular," it was "wonderful." They're "glad to be Americans." Even among those who are not Americans there is a sense of pride at what happened here.*

Broun had watched with several thousand people at an outdoor viewing area at the Tomorrowland section of Disneyland. He described the moment of touchdown to Cronkite.

Broun: *Well, usually, Walter, at moments in history people are dressed accordingly. You see those paintings in which everybody is in uniforms, and medals or crowns or whatnot. There aren't the costumes of destiny here at Disneyland, but at the moment when the landing actually took place, we had a huge throng spread all back here and on the roofs of the buildings nearby, all looking at the screen. And at the moment that you announced it was down, there was kind of a long burst which was partly applause of praise, and partly, I think, applause of relief. Normally this is a happy, jolly place, but it became as tense as an operating theatre during those last few seconds. [To spectator] Now you were here right through it, weren't you? What did you think watching it come down?*

Man: *Well, it was absolutely exciting. There is no doubt about it. I was not born here, but this is one moment I want to share with the American people. Because they have been frontier people all through the history of mankind, and this is one frontier which shows the absolutely amazing way they could conquer. And I congratulate the whole world, especially the United States, for this discovery and achievement.*

And at the Smithsonian in Washington, the same excitement and tension was evident in a conversation Roger Mudd had with a tourist.

Mudd: *Did you think we'd make it?*

Man: *Yes sir, all the way.*

Mudd: *Was there ever any doubt in your mind about it?*

Man: *No sir, not really. I knew we could do it.*

Mudd: *No knot in your stomach?*

Man: *Oh, yes sir! Yes sir!*

The man's last reaction seemed to express the feelings of the world's people that afternoon. Everyone knew they would do it, but the concern and tension in the last few seconds were unbearable. Eric Sevareid had joined Cronkite to discuss those last few seconds and the significance of the moment in history.

Cronkite: *Eric Sevareid, I guess this day has given us our biggest story. We're contemporaries, and we've covered World War II together, and the conflicts since then and the comings and goings of heads of state. But, I don't think anything compares with this.*

Sevareid: *I think, Walter, that it's easier to be an active participant in world-shaking things like this than an observer who can only sit. At least it's easier on the nerve ends. The thing that got us all downstairs in another office, watching this, in those last few moments and seconds, was the steadiness of those voices of the two men. That lean prose of Armstrong, with not a wasted word. He said precisely the words and the figures that he had to recite, that had to be known. No more. In an absolutely dead-calm voice. And yet that prosaic conversation, I suppose those words—ordinary as they sounded—are going to be reread and relistened to for maybe a hundred years. And as an old-fashioned humanist, it seemed to me a little reassuring that in those last seconds the human hand and eye had to take over from the computers, if I understood exactly what was going on in that Lunar Module. I gather that if it had all been computerized to the last second, we might have wrecked this "Santa Maria" on a rock.*

Of course, the moon now is something different for the whole human race. There's a price for everything. There isn't any gain without some loss. When you got the telephone you lost privacy. When we got the airplane, we lost the sense of travel. Just translated from one condition to another. When we physically possess the moon, I suppose it will dawn on us, in a sense, that spiritually we lose the moon we had for thousands of years. At least in terms of its remoteness, its wonder and mystery, the romance and the poetry of it. There's a line in Shakespeare's Henry IV, *Walter, that says, "methinks 'twere an easy leap to pluck bright honor from the pale-faced moon." And how any actor henceforth can utter that with a straight face, I don't quite see.*

As Dr. Paine said earlier, on the earlier Apollo flights that this was a "triumph for the squares." And so is this. I suppose that's one reason that the only people who seem a little blasé and find this a little distasteful are the young intellectuals of a moral and sociological bent. And this is being done, of course, by older people, and by the State machinery that they don't like very much. I think it's really because people, as the Frenchman said long ago—"always prefer what is not to what is." But the reaction of those people we saw on the beaches, and the airports, ordinary people—this is the governing reaction. They do like life. They do like reality. They are excited. I think that's the normal healthy heart of the country, and mind of the country, that you're hearing then.

Cronkite: *I was rather appalled as a supposedly professional reporter—communicator—talker—that sitting here, I really was caught speechless. I don't think that's happened in my life. Perhaps the moment when we knew that President Kennedy was dead. But this moment we were prepared for really. We've been talking about what these fellows were going to say when they landed, and they were talking away. They had their things to report, their job to do. But I—my mouth and my throat...*

Sevareid: *Sometimes, silence is the most eloquent thing on earth. Silence.*

Cronkite: *As a matter of fact, Armstrong's words were as eloquent as you could ask: "Eagle has landed!" What more could you say?*

The announcement that most people in the control room had been expecting came at 6:15 p.m. EDT. Armstrong would open Eagle's hatch to begin his moon walk at nine o'clock, five hours earlier than planned.

This was going to raise havoc with the broadcast plans for the early Sunday evening period. This was the time when Wussler, Miss Richman and Cross had planned to present Cronkite's exclusive film interview with Lyndon Johnson, the Orson Welles-narrated science fiction piece and a Mike Wallace–Orson Welles interview in London, panel discussions on the significance of the day and a number of the other films specially prepared for the broadcast. With the word that the astronauts' EVA (Extra-Vehicular Activity) had been moved ahead five hours, these plans were postponed.

Two hours after her husband had landed on the moon, Mrs. Neil Armstrong stepped to the porch of her Houston home to greet friends, neighbors and the press. In reply to one question she admitted that her husband had taken something to the moon for her, but refused to reveal what it was. Cronkite had learned an interesting sidelight about one of the personal things the astronauts had taken along with them.

Cronkite: *While we heard a minute ago from Janet Armstrong that Neil had taken a little packet of something memorable for her along with him, we know that the others have as well. And also they're being secretive about what it is, as all of the astronauts have from the first flight of Alan Shepard in 1961. Buzz Aldrin did take something most unusual with him today, and it has become public. Made public by the pastor at his church outside of Houston. He took part of the Communion Bread Loaf, so that during his evening meal tonight he will, in a sense, share communion with the people of his church, by having a bit of that bread up there on the surface of the moon. The first Communion on the moon.*

The satellite was in operation and reaction stories began to come in from around the world. Mike Wallace, who had watched the landing at the residence of British Prime Minister Harold Wilson, was the first to report.

Wallace: *The Prime Minister said that he had watched the account of the last half hour before Apollo's touchdown, and when it was determined that astronauts Armstrong and Aldrin were safe and all systems "GO," he came out of his study to make this statement:*

Wilson: *I think the first feeling of everyone in Britain who has seen and heard the news of the incredible achievement will be one of heartfelt relief that this very dangerous part of the mission has been safely accomplished. We all know there's still more to be seen through. And I think the second feeling will be one of tremendous admiration. Admiration first for the way in which this great and historic achievement was conceived and planned. I suppose it's an achievement which incorporates all of the work, all of the discoveries of the mathematicians, and the scientists and the space experts, almost from the earliest days of mathematics and science. An incorporation that acknowledges the experience of many nations. The final credit obviously goes to those in the United States, who have developed the missiles, who have developed the technology, who have developed the modules, who have developed the computerization. And to the overall perfection of planning of the United States Space Agency. With equal admiration for the courage and fortitude not only of these three gallant astronauts, but all those others who have gone before and who have made this possible. Admiration, too, for the bravery and the fortitude of their families, who have been unable to do more than stand by and watch. And it must be really a tremendous sense of pride at this moment.*

Wallace: *Afterwards the Prime Minister told us that he wants to send a message of congratulations to President Nixon, but that he will wait until the Eagle and Columbia are joined together again. Incidentally, Walter, I asked the Prime Minister if he had any idea of what Russia's Luna 15 was up to. He joked . . . he joked, mind you, that five minutes before Apollo's touchdown, he had fully expected Luna to land half a dozen Russians on the moon, Leonid Brezhnev among them. And one of his advisors said: "Yes, and only five of them will return to earth!"*

CBS Newsman Bob Simon had been at London's Trafalgar Square where, with thousands of Britons and tourists, he watched the landing on a huge Eidophor screen.

Simon: *There were no shouts of ecstasy here the moment of touchdown, no screams, no cries, no yelling. It was mainly a big sigh of relief. Peo-*

ple here have been feeling this very deeply. And the main reaction after touchdown was "My God, could it really happen so easy?"

Daniel Schorr was standing by in Amsterdam to give the Dutch reaction to the landing.

Schorr: *It was all a little bit different here in Holland. There were no mass demonstrations in any squares. We don't have the Dutch Prime Minister available to speak to you. One of the reasons is that the Dutch are a more careful people. Even Prince Bernhard, whom we'd hoped to have here today, said he would like to see the whole thing a success before he issued a statement. I think it's in the cautious Dutch character to wait until everything is completely successful before you get any official statements. For the Dutch, the grand climax will be when they see an American man on the moon.*

Peter Kalischer was in Paris, where the reactions of the people on the streets of that city well described the excitement felt at the moment of touchdown.

Woman: *I think this is a very important day and a wonderful day as far as the whole universe is concerned, because it's the landing on the moon.*

Man: *I think it's just wonderful to be on earth, and to learn what's going on on the moon tonight. Pray God for these people, and I thank the Americans also for what they have done for the world. I am very proud of the Americans.*

The Italians were ecstatic on learning that Armstrong and Aldrin had safely landed on the moon. Winston Burdett was reporting from Rome.

Burdett: *Pope Paul watched television tonight at the Vatican Observatory. And at the moment of landing, he raised his hand in blessing and said: "Honor, greetings and blessings to you, conquerors of the moon." The headline in Rome's leading newspapers is a single word that fills half the front page—"Landed." No other event that I can recall has so* 87 *fired the Italian imagination. There is unbounded wonder at the achievement, at the capacities, the power, the perfection of American technology. And in the press, there is an outpouring of superlatives . . . "prodigious," "stupendous." If there is one constant theme in Italian comment it is this: the United States by an unparalleled coordination of a vast number of technologies, has overtaken and surpassed the Russians in their own field of collective organization. It is a pretty emotional experience for Italy. For the moment at least, all doubts and questions were swept away in the euphoria. If man can reach the moon, one paper says, "then there are no material problems that cannot be solved in human society on earth."*

For the first time people in Belgrade, Yugoslavia, were seeing full live coverage of an American manned space flight. Their enthusiasm for the venture would be hard to match, as Bill McLaughlin reported from the Yugoslav capital.

McLaughlin: *The flight of Apollo 11 is being followed here with every bit as much interest as in the United States. Yugoslavia in fact has adopted the three American astronauts as its own heroes. Here, as in most of the world, Apollo 11 is being considered as "man's greatest adventure."*

An expert and most interested observer sat with Morley Safer at the Jodrell Bank observatory in Manchester, England. Sir Bernard Lovell, the world's foremost radio astronomer, had waited many years for this moment.

Safer: *Sir Bernard, just about everybody here was fascinated when we heard that cool talk-down. It must have been a very special moment for you, sir.*

Lovell: *It was certainly tremendous. As an ordinary individual, I share the emotions of most other people who listen to these moments of drama. As a scientist, I feel that a completely new avenue, a revolutionary*

88

one, has been opened for our further understanding of the universe. As for the Americans, it is a very, very great moment indeed. By this landing of two men in this extraordinarily efficient and calm way they have demonstrated the superiority in a manner in which we feel it exceedingly difficult to comprehend. And I think that it is very hard to find words to express one's admiration, and one's congratulations to our American colleagues on this very dramatic achievement.

The world was waiting for Neil Armstrong to back through the open hatch of the Lunar Module, and down the ladder to the surface of the moon.

There had been other interesting reactions to the landing. Cronkite reported that at New York's Yankee Stadium 35,000 people watching a baseball game had cheered so loudly that the players got involved and called for a moment of silent prayer of thanks for the successful landing and the astronauts' safety. A newborn baby in Beirut, Lebanon had been named "Apollo," and another baby had been named Neil Michael Edwin, in honor of Armstrong, Collins and Aldrin.

Frank Borman, commander of the Apollo 8 mission, the first to leave earth orbit on its historic Christmas Eve circumlunar flight, had been with President Nixon shortly after the touchdown. Later, he was at the Smithsonian Institution with Mudd and White House Correspondent Dan Rather.

Mudd: *Colonel, it's good to see you and have the opportunity to talk with you. Dan is here from the White House, because that's where you spent part of your day. I'd like just to ask you what you did at the White House and why you were there?*

Borman: *Well, Dr. Paine, who runs NASA, appointed me a sort of a liaison officer with the White House for the Apollo 11 activities, and I've been coming up here for—oh, a month, two or three days a week, during preparations and briefing and so on.*

89

Mudd: *Did you serve as the space interpreter for the President during the technical periods of the voyage?*

Borman: *That's right.*

Mudd: *When were you with the President?*

Borman: *Well, I was with the President immediately after the landing, and I was with him during the launch. We just came from quite a discussion with him. I think he was impressed by the feat as were most of the other people who were watching it.*

Rather: *As a matter of fact, Colonel, we have a quote here from the President that he gave out to the White House press office, in which he said he was "glued to the television set." I'm happy to report, by the way, Walter, that the White House said that the President watched the moon landing on CBS. He only had one set; it was a small portable color set in the Executive Office Building. "It excelled anything that I've ever seen," said President Nixon. And he said that those last 22 seconds before landing were the longest he'd ever lived through. Did he have any specific questions for you, Colonel, once you got in the room with him?*

Borman: *Yes, we talked about the landing, and how Neil could take over, and do it manually. Obviously from the description I heard Neil give, if it hadn't been for the fact that he was able to maneuver it at last, it might have been a little different story. We also got into some of the philosophic aspects of the space flight. And I reminded the President that I had read in a periodical not long ago that he made one of the first statements favoring space in this country back in 1957 in San Francisco. He informed me that he remains a space activist and he's all for this country having a vital and forward-thrusting space program.*

Correspondent Bill Plante had spent most of a rainy day on the streets of New York looking for the reactions of people on the day man landed on the moon. At the time of the landing, Plante and a mobile unit were in Mount Morris Park where nearly 50,000 people were attending a Soul Festival of the Harlem Cultural Festival. Plante asked some of the people at the Soul Festival for their reactions to the safe landing.

Man: *I think it's very important, you know. I think that's the greatest feat man ever accomplished, so far, going to the moon. But I don't think it's any more relevant, than, you know, the Harlem Cultural Festival here. I think it's equal.*

Man: *It's practically wasted, as far as I'm concerned, in getting to the moon. It could have been used to feed poor black people in Harlem and all over the place, all over this country. You know, like never mind the moon, let's get some cash in Harlem.*

The enthusiasm for this day in history was not unanimous. There were voices of dissent, as Cronkite reported, among the youth and Blacks in the United States, and complaints from abroad that "priorities have been a little confused." But as Cronkite also noted, "By and large, it's been a day of rejoicing around the world, for the success of this moon mission."

It was 7:00 p.m. EDT, and the moon walk was scheduled to begin in two hours.

Vice President Agnew had arrived at the Smithsonian Institution, where he repeated to Roger Mudd his belief that America's manned space program should have Mars as its next target.

Mudd: *Were you surprised at the reaction on Capitol Hill to your advocacy of a trip to Mars by the end of this century?*

Agnew: *Not really. I suppose their reaction was that of people who are deeply concerned about the domestic problem and the drain of funds for it. I said what I said not representing the Administration but as an individual, but I want to make one thing completely clear. I'm totally confident, through my conversations with the President, that he is a space activist. And when he receives the recommendations of the task force that I now head, he will make a very far-reaching and ambitious*

91

decision on the space program. I don't know just what it will be. I know he wants to leave his options open till he's heard from everybody. But don't expect a bland, conservative decision out of President Nixon because after all it was President Nixon, right after Sputnik in 1957, who was the first public official to speak out, when some people were playing down the appropriations of that effort, and indicated that he was very much concerned about the U. S. not losing the leadership in space.

Neil Armstrong's walk on the moon was the culmination of the dreams of many men. One of them, Dr. William Pickering, Director of the Jet Propulsion Laboratory in Pasadena, California, had been involved in the very earliest unmanned probes of the moon. In the next two weeks he would be seeing the results of two unmanned probes of Mars, the planet that clearly was on the minds of a great many people that day. Dr. Pickering was with Bill Stout in California.

Stout: *You know there's been an argument for years about manned versus unmanned, which is more efficient and more economical, and really it's an argument that's almost resolved today, with the mystery of Luna 15, circling the moon at the same time that Apollo is there. Does it seem to you, sir, that that whole dispute has become completely irrelevant, with that landing?*

Pickering: *I think it's irrelevant, because there are always things for man to do and there are things for instruments to do. The instruments of the early unmanned probes have paved the way for Apollo. They've shown much—they've shown us how to navigate to the moon and what the surface of the moon actually looks like. We've got a good idea of what it's going to be like when Armstrong first steps out on the moon in a few minutes. And this has all been an essential part of the total development. As we look ahead, we can see that the exploration of the moon will require both manned and unmanned devices. There are going to be parts of the moon which will be very difficult to get to with a manned vehicle. And these, we expect, will be explored with unmanned devices. Man, however, will certainly be a key part of the exploration. When we talk about unmanned spacecraft we have to remem-*

92

ber that there really is a man involved. The man may be sitting down here on the earth instead of in the spacecraft, but the sorts of devices which are being flown nowadays are complex devices, with command capabilities, with control capabilities, and a man must some way be making those decisions. And so we have manned and unmanned devices working together to explore space.

The rejoicing of this day had an edge of sadness. Armstrong and Aldrin carried the memories of five men, three American and two Russian, with them to the surface of the moon. The men, three astronauts and two cosmonauts, had died during training or in actual space flight.

Cronkite: *The three Americans, Gus Grissom, Ed White and Roger Chaffee, died in a fire, as you know, on the launching pad in Florida as they were rehearsing for their flight. That was in January 1967. And the two Russians who have died, Komarov, whose parachute fouled as he returned from a space flight, and Yuri Gagarin, the first man in space who died in a training airplane accident. Medals have gone to the moon, some of them actual Gagarin and Komarov medals given to Frank Borman when he was over in Russia not very long ago. And three mementos for Grissom, White and Chaffee. And the courage of the families of the three American astronauts who died in preparing for the success of this mission today, and contributed to it, with their previous space flights, and with the preparations they had made up to the time of the fatal fire, is remarkable.*
And particularly, if any could be singled out, I think it would be the White family. Ed White was our first space walker. His father was a general in the Air Force who went right out and spoke for the space program around the United States and continued to impress people with the necessity for the program to go on—how his son would have wished for it to go on. Today, Pat White, widow of Ed White, was one of the first neighbors to rush into Neil Armstrong's home, outside of Houston at the Space Center. She still lives there with the White children. She offered her smiling, happy congratulations. She said, "I

93

couldn't be happier. It's the culmination of a dream, a culmination for a lot of people. Ed would have been thrilled to see this day come."

A strange, almost tribalistic, ritual was in progress in New York's Central Park in the minutes before Neil Armstrong set foot on the moon. Several thousand people, standing in the rain, were attending what New York City had billed as a "Moon-In" and Bill Plante was there with the "Flash Unit." CBS News had erected a gigantic Eido-phor screen which had attracted most of the people in the Park.

The heavy rain had made it nearly impossible to work in the Park. People were standing in three or four inches of mud to watch the moon walk. "We used to blame bad weather on the atomic bomb," joked CBS News producer David Fox, "but after sloshing around New York for a day, I was convinced that it had something to do with men landing on the moon."

The people's reasons for being there were pretty much the same.

Woman: *Well, it was something that we wanted to see together with a lot of people.*

Man: *Well, it's been a little hard, but I think it's kind of fun out here. Who'd want to sit back in a stupid old room.*

Cronkite: *It doesn't matter where a man is, whether he's back in that stupid room, as the fellow said, or whether he is sharing this experience as that girl said, with a lot of others as she feels they should.*

Cronkite and Schirra sat at the anchor desk during the final minutes before Neil Armstrong was to leave the Lunar Module. It was now 10:25 p.m. EDT, and the astronauts were nearly one hour and a half behind schedule for their Extra-Vehicular Activity (EVA).

Cronkite held a copy of the first edition of Monday morning's *New York Times*. In boldface, the headline read, MEN LAND ON THE

MOON. Never before had the *Times* run a headline in such large type. As Cronkite said, "a record breaker, another record breaker today, but fitting the occasion."

There was a pervading sense that a new era was dawning.

"Okay, Houston, I'm on the porch!"

Neil Armstrong had slowly backed out of the hatch of the Lunar Module and was on his way down the ladder.

Houston: Roger. We copy and we're standing by for your TV.

Armstrong: Houston, this is Neil, radio check.

Capcom: Neil, this is Houston. You're loud and clear. Break, break. Buzz, this is Houston. Radio check and verify TV circuit breaker in.

Aldrin: Roger, TV circuit breaker's in. Receive loud and clear.

Capcom: Man, we're getting a picture on the TV.

As at touchdown, a cheer rolled through the control room as a hazy, indistinct picture flashed on the screen. At first the picture was upside down, but Houston quickly converted the image. The shadowy figure gingerly moved down the ladder rung by rung.

Capcom: Okay, Neil, we can see you coming down the ladder now.

Armstrong: Okay, I just checked getting back up to that first step, Buzz. It's not even collapsed too far, but it's adequate to get back up.

Capcom: Roger, we copy.

Armstrong: It takes a pretty good little jump. I'm at the foot of the ladder. The LM footpads are only buried in the surface about one or two inches, although the surface appears to be very, very fine grained as you get close to it. It's almost like a powder. Now and then it's very fine. I'm going to step off the LM now.

Cronkite: *Boy! Look at those pictures! It's a little shadowy, but he said* *he expected that in the shadow of the Lunar Module. Armstrong is on the moon! Neil Armstrong, a 38-year-old American standing on the surface of the moon! On this July twentieth, nineteen hundred and sixty-nine.*

Armstrong: **That's one small step for a man, one giant leap for mankind!**

It was 10:56:20 p.m. EDT, Sunday, July 20, 1969.

Armstrong: **The surface is fine and powdery. I can pick it up loosely with my toe. It does adhere in fine layers like powdered charcoal to the sole and sides of my boots. I only go in a small fraction of an inch. Maybe an eighth of an inch, but I can see the footprints of my boots and the treads in the fine, sandy particles.**

Schirra: *Oh, thank you television for letting us watch this one!*

Cronkite: *Isn't this something! 238,000 miles out there on the moon, and we're seeing this.*

Armstrong: **There seems to be no difficulty in moving around as we suspected. It's even perhaps easier than the simulations at one-sixth G that we performed on the ground. It's actually no trouble to walk around.**

Cronkite: *Gee, that's good news!*

Armstrong: **The descent engine did not leave a crater of any size. There's about one foot clearance on the ground. We're essentially on a very level place here. I can see some evidence of rays emanating from the descent engine, but very insignificant amount.**

Aldrin had started to pass a still photograph camera down to Armstrong.

Armstrong: **It's quite dark here in the shadow and a little hard for** me to see that I have good footing. I'll work my way over into the sunlight here, without looking directly into the sun. Looking up at the LM, I'm standing directly in the shadow now, looking up at Buzz in the window. And I can see everything quite clearly. The light is sufficiently bright, back-lighted into the front of the LM, that everything is very clearly visible.

Armstrong had taken his Hasselblad camera and had backed away from the LM for the first of a series of photographs. That was a little out of step with the flight plan. He was supposed to take a contingency sample of rocks first, and after he had been at the picture-taking for a while, Houston asked if he had "copied" about the contingency samples.

Then it was Aldrin's turn. The voice from inside Eagle politely made an inquiry.

Aldrin: Okay, going to get the contingency sample now, Neil?

Armstrong: Right.

Cronkite: *Now Aldrin is bugging. The Commander is going to get tired of those guys in a minute, and get them a contingency sample just to quiet them. Nag…nag…nag…*

Armstrong's shadowy figure moved, almost bounced, away from the Lunar Module to get the rock samples.

Armstrong: This is very interesting. It's a very soft surface, but here and there where I plug with the contingency sample collector, I run into a very hard surface. But it appears to be very cohesive material of the same sort. I'll try to get a rock in here. Here's a couple.

Having done his duty, Armstrong returned to his inspection of the lunar surface. He described it to the world.

Armstrong: It has a stark beauty all its own. It's like much of the high desert of the United States. It's different but it's very pretty out here.

Then it was time for Buzz Aldrin to make his trip down the ladder. Armstrong guided him down, making sure that Aldrin's PLSS (Portable Life Support System) backpack cleared the hatch on the way out. Then Buzz was on the porch.

Aldrin: **Now, I want to back up and partially close the hatch, making sure not to lock it on my way out.**

Armstrong: **A good thought.**

Aldrin: **That's our home for the next couple of hours and I want to take good care of it. Okay, I'm on the top step and I can look down over the landing gear pads. That's a very simple matter to hop down from one step to the next.**

Then Aldrin had joined Armstrong on the lunar surface, the second man to walk on the moon.

Aldrin: **Beautiful, beautiful.**

Armstrong: **Isn't that something. Magnificent sight down here.**

Aldrin: **Magnificent desolation.**

Armstrong and Aldrin inspected the exterior of Eagle and Aldrin had his first close look at the surface. Then as Armstrong prepared to move the television camera away from the LM, the two astronauts moved to the stainless steel plaque on the front landing gear.

Armstrong: **For those who haven't read the plaque, we'll read the plaque that's on the front landing gear of this LM. First there are two hemispheres, one showing each of the two hemispheres of the earth. Underneath it says, "HERE MEN FROM THE PLANET EARTH FIRST SET FOOT UPON THE MOON. JULY 1969 A. D. WE CAME IN PEACE FOR ALL MANKIND." It has the crew members' signatures and the signature of the President of the United States.**

Cronkite: *Oh, boy!*

Armstrong moved the camera some 50 feet away from Eagle. Before he set it in its permanent position he panned the lunar landscape in the vicinity of the LM.

Schirra: *I sure hope there's no area in this world that's blacked out from television right now.*

Cronkite: *But there is. More than a fourth of the people of the world are being denied this picture by their rulers. Through most of the world, though, including the Communist countries of Eastern Europe, the picture is being shown.*

With the camera in place, the astronauts got to some of the more serious tasks of their mission on the moon. Aldrin erected the solar wind experiment, which would measure the flow of protons from the sun on the lunar surface, and the two astronauts began taking the bulk rock samples so anxiously awaited by the astrogeologists on earth.

In the meantime, Columbia had come around to the near side of the moon again and CAPCOM in Houston described what was happening 69 miles below Mike Collins on the moon's surface.

Capcom: **Columbia. This is Houston reading you loud and clear. Over.**

Columbia: **Yes. This is history. Yes. Read you loud and clear. How's it going?**

Capcom: **Roger. The EVA is progressing beautifully. I believe they are setting up the flag now. I guess you're about the only person around that doesn't have TV coverage of the scene.**

Columbia: **That's all right. I don't mind a bit.**

Then Armstrong and Aldrin planted the American flag on the lunar surface. Houston informed Collins that the flag was in place.

Capcom: Columbia. Yes, indeed, they've got the flag up and you can see the Stars and Stripes on the lunar surface.

Collins: Beautiful. Just beautiful!

Cronkite: *There it is, a little U.S. flag on the surface of the moon!*

Then Aldrin explained some of the problems they were having moving around on the lunar surface.

Aldrin: You do have to be rather careful to keep track of where your center of mass is. Sometimes it takes about two or three paces to make sure you've got your feet underneath you. About two or three or maybe four easy paces can bring you to a nearly smooth stop. Next direction – like a football player you just have to split out to the side and cut a little bit. One called a "kangaroo hop" does work, but it seems that your forward ability is not quite as good.

Aldrin demonstrated the "kangaroo hop," taking long hops in and out of camera range to the delight of Cronkite and Schirra.

Schirra: *Look at the powder come up there.*

Cronkite: *They're beginning to get pretty frisky up there.*

Schirra: *Oh, beautiful!*

Capcom: Tranquility Base, this is Houston. Could we get both of you on the camera for a minute please?

Capcom: Neil and Buzz, the President of the United States is in his office and would like to say a few words to you.

Armstrong: That would be an honor.

Capcom: Go ahead, Mr. President, this is Houston. Out.

One half of the split screen showed the President sitting at his desk. The other half showed Armstrong and Aldrin standing motionless on either side of the flag.

Nixon: Hello, Neil and Buzz, I'm talking to you by telephone from the Oval Room at the White House, and this certainly has to be the most historic telephone call ever made from the White House. I just can't tell you how proud we all are of what you have done. For every American this has to be the proudest day of our lives. And for people all over the world, I am sure that they too join with Americans in recognizing what an immense feat this is. Because of what you have done the heavens have become a part of man's world. And as you talk to us from the Sea of Tranquility, it inspires us to double our efforts to bring peace and tranquility to earth. For one priceless moment in the whole history of man, all the peoples on this earth are truly one. One in their pride in what you have done, and one in our prayers that you will return safely to earth.

Armstrong: Thank you, Mr. President. It's a great honor and privilege for us to be here representing not only the United States, but men of peace – of all nations – with an interest and a curiosity and a vision for the future. It's an honor for us to be able to participate here today.

Nixon: And I thank you very much, and I look forward, all of us look forward, to seeing you on the *Hornet* on Thursday.

Armstrong: Thank you.

Aldrin: I look forward to that very much, sir.

Then Armstrong and Aldrin saluted. The astronauts then turned from the camera to continue their work at Tranquility Base.

Armstrong finished taking the bulk rock sample, while Aldrin photographed the LM from all sides for a record of its condition after the landing.

The remaining hour and a half of the EVA would be spent photographing and describing the terrain, and deploying the package of scientific experiments they had brought to leave on the moon. Two of the more important experiments they would leave on the surface were a laser reflector, that would allow scientists to bounce a laser beam off the moon and back to earth, and a seismic device that would measure and record everything from an earthquake, or moonquake, to a meteor

101

striking the surface. It would also record the landing of a manned or unmanned spacecraft on the surface.

In the waning minutes of the moon walk, Bob Wussler had left his place in the control room to visit the guest reception area. There, producer Burton Benjamin had been receiving guests originally scheduled to appear that evening. When the EVA had been moved forward the guests stayed on. Wussler recalls that he walked into what "had to be the best party in town." Arthur C. Clarke, Buckminster Fuller, Keir Dullea, Reverend Andrew Young, Dr. James Watson and Kurt Vonnegut, Jr. were only a few of the guests who had remained to watch the moon walk. "There was an air of jubilation in the room" says Wussler. "The guests and CBS, CBS News and CBS Television Network executives who had gathered there were enthralled by the pictures they were seeing from the moon."

CBS executives had another cause for satisfaction, a very down-to-earth one. The early National Arbitron audience ratings for the period up to 6:00 p.m. EDT, Sunday had come in, and CBS News' television coverage had earned 44 percent of the viewing audience, compared to NBC's 32 percent and ABC's 14 percent—a margin for CBS News which was later to widen even further. It was a tremendous psychological boost at a time, over thirteen hours into the broadcast, when everyone was starting to show the wear of the long day.

At 12:57 a.m. EDT, on Monday, July 21, a reluctant Buzz Aldrin, after much prodding and a couple of direct orders from Houston, returned to the LM. From there he helped Armstrong maneuver back into Eagle the rock samples and film that would return to earth.

Then at 1:09 a.m. EDT, Armstrong began to climb the ladder of the Lunar Module. In two bounds he was on the porch, and at 1:11 a.m., the hatch door closed behind him. An epic two hours and 31 minutes had come to an end. Eagle sat silently on the atmosphereless, windless moon.

These were moments of silence. There was no communication from

Armstrong and Aldrin in the Lunar Module. There would be none until they were out of their space suits and using the Lunar Module's communications system. In many ways they were welcome moments after the excitement of the past 14 hours. One could sit back and reflect, as Cronkite and Eric Sevareid did, on what they had seen and experienced.

Cronkite: *Man has landed and man has taken his first steps. What is there to add to that?*

Sevareid: *At this hour one can only subtract. I don't know what one can add now. We've seen some kind of a "birth" here. And I'm sure that to many, many people the first scene of Armstrong emerging, must have seemed like a birth. One's image of this clumsy creature, half-blind, maneuvering with great awkwardness at first, and slowly learning to use its legs, until, in a rather short time, it's running.*

And in this new world, this new reality. And the quickness of the adjustment of the human body, and the nervous system. The weight of gravity on earth. Just the other day they were at the Cape; then weightlessness of several days; and then to the moon's one-sixth gravity. And somehow the body adjusts with that speed and in totally different elements. This is what overwhelms you. And Armstrong's words. He sounded very laconic, unemotional. His mother said as she heard them on the air that she knew that he was thrilled. And I think we'd have to take a mother's word for that. And then when they moved around, you sensed their feeling of joy up there. I never expected to see them bound, did you? Everything we've been told was that they would move with great care. Foot after foot with great deliberation. We were told they might fall. And here they were, like children playing hopscotch.

Cronkite: *Like colts almost.*

Sevareid: *Like colts finding their legs, exactly. I must say, as somebody who loves the English language, I have such a great gratitude that the first voices that came from another celestial body were in the English tongue, which I feel is the richest language of all. I think it is the greatest vocabulary. And maybe only 300 million people or so on* *the earth speak and understand it. But I never expected to hear that word "pretty." He said it was "pretty." What we thought was cold and desolate and forbidding—somehow they found a strange beauty there that I suppose they can never really describe to us. So we'll never know.*

Cronkite: *It may not be a beauty that one can pass on to future beholders, either. These first men on the moon can see something that men who follow will miss. Just a smidgin of it anyway.*

Sevareid: *We're always going to feel, somehow, strangers to these men. They will, in effect, be a bit stranger, even to their own wives and children. Disappeared into another life that we can't follow. I wonder what their life will be like, now. The moon has treated them well, so far. How people on this earth will treat these men, the rest of their lives, that gives me more foreboding, I think, than anything else.*

Reporters had talked to Neil Armstrong's parents, Mr. and Mrs. Stephen Armstrong, at their Wapakoneta, Ohio home, a few minutes after their son returned to Eagle.

Mrs. Armstrong: *I was afraid the floor of the moon was going to be so unsafe for them. I was worried that they might sink in too deep. But no, they didn't. So it was wonderful.*

Reporter: *Mr. Armstrong, what were your feelings?*

Mr. Armstrong: *I was really encouraged the way I understood that Neil had guided the craft to another area. And that would signify that the original was not exactly as they had planned.*

Reporter: *What about his voice? Did it sound any different? Or did it sound calm and normal to you?*

Mrs. Armstrong: *I could tell that he was pleased and tickled and thrilled. He was much like he always has been.*

Mr. Armstrong: *I had the same feeling, that it was the same old Neil.*

Cronkite: *That brief stirring report from the home of the parents of Neil Armstrong, Mr. and Mrs. Stephen Armstrong. And Neil Arm-*

strong's wife was reported to have said, "I can't believe it's really happening" when the first step was taken, and when he took that step, she said, "That's the big step!" and then she had a little wifely advice. She said, "Be descriptive now, Neil," as he began to walk upon the moon, probably heeding the suggestions of a lot of reporters, including this one, that perhaps he might talk a little bit more than he had on the rather taciturn trip out to the moon. We have no complaints now, none at all, certainly about this great feat by these two very great fellows who got their feet on the moon, and Mike Collins who made it possible by playing watchdog up above them there.

The first men to see the moon at close hand were the Apollo 8 crew members, Frank Borman, James Lovell and William Anders. If Apollo 8 had been the moon landing flight, Anders, recently appointed executive secretary of the National Space Council, would have been the Lunar Module pilot, the Buzz Aldrin of an historic day in space history. Anders had watched the walk at Mission Control in Houston, and after the hatch closed, he joined Bruce Morton in the CBS News studio at the Manned Spacecraft Center to describe what Apollo 11's accomplishments meant to him.

Morton: *Eric Sevareid, a little while ago, mentioned that Neil Armstrong had used the word "pretty" to describe the moon. Did that occur to you when you got the closest look anyone had had?*

Anders: *No. I think magnificent would be a good word. But I think somebody said "magnificent desolation." To me the earth was pretty. But I didn't really think the moon was.*

Morton: *Did you think during the Apollo 8 flight, that this would be happening this soon?*

Anders: *Yes. I did. We had the timetable pretty well laid out. We knew there had to be quite a bit of success in the program, but then people had worked very hard to get pre-planned for this success, and I thought it would come to pass.*

105

Morton: *What, if anything, surprised you about the landing or the walk?*

Anders: *I was surprised by the great mobility that the crew had. I thought that they wouldn't be able to move without difficulty. But I didn't really feel that they would have that much freedom of motion.*

Morton: *Almost frisky. Bouncing about.*

Anders: *They were moving on the moon like a bunch of Texas jackrabbits up there. The other thing that surprised me was that apparently the struts of the vehicle didn't compress as much as I had planned and expected. I guess that just attests to Neil's ability to make a nice smooth landing.*

Morton: *Is that the toughest part of the flight, do you suppose?*

Anders: *Well, that's the part that I was watching the closest, the actual touchdown itself. And that's one of the few things we had quite a bit of trouble simulating here on earth, and it's one of the things we'd never done before.*

Morton: *Looking back over this day, what do you think you'll remember most? The landing, the first foot coming down?*

Anders: *Well, certainly the first foot coming down with the television coverage was a very remarkable thing. But I think the landing itself will be the thing I'll remember the most.*

Morton: *Spoken like a pilot.*

Arthur C. Clarke and Robert Heinlein, author of *Destination Moon*, were together following the moon walk. Heinlein's movie, made in 1950, had visualized many of the same things that had been seen during the two hours and 31 minutes when Armstrong and Aldrin were on the surface. In fact, the pictures from the moon had, as Cronkite remarked earlier, "a science fiction movie quality" about them. Clarke and Heinlein discussed with Cronkite the future of space.

Heinlein: *I'm not at all sure just how things are going to be done. I'm simply certain that they're going to be done in a big way, that we're going out to all of the planets. That we're going out to the stars. We're going out indefinitely. There's just one equation that everybody knows: E=MC². It proves the potentiality whereby man can live anywhere where there is mass. He doesn't have to have any other requirement but mass, with the technology that we now have. And this human race will do so.*

Clarke: *And when they do go out, just as when they came to this country, they'll forget their original nationalities. We're going to see them going out into space as nations which will develop new ideas. And as far as this country is concerned, I do hope that this great lifting of the spirit which we've all experienced today will make a change in morale and help this country get away from the defeatism of the past.*

Heinlein: *I do hope so. There have been too many of the young people in this country who have the defeatist attitude toward things and I hope that this will give them the "lift," the "esprit de corps," to realize how terribly important this is. Not to them alone, but to their children, their grandchildren, through thousands of years. This is it. This is the great day.*

Apollo 11 had been television's story. Hundreds of millions of people around the world had seen Neil Armstrong's foot touch the lunar surface at the precise moment he left the Lunar Module. Together with the miracles of film and videotape, the feats of Armstrong and Aldrin would depend upon the printed word for permanence. Monday morning's newspapers would be the keepsakes, the record for years to come of the historic date. Less than two hours after the walk was completed, the New York *Daily News* was about to roll the presses on a paper bannering the headline "MEN WALK ON THE MOON." Correspondent Bill Plante was there.

Plante: *The presses at the* Daily News *are running about two hours late* *tonight with the Four Star Edition. But, of course, this is no ordinary night. The headline of the Four Star reads "MEN WALK ON THE MOON." And that of course, is the big story, the story that they have been calling the biggest in the history of mankind. Print has a certain permanency greater than our fleeting words, a permanency that varies from one man to another. For some it will last as long as it takes to wrap tomorrow's garbage. For others it will be until the newsprint begins to turn yellow and shrivel with age, and becomes dust in someone's closet.*

3:15 a.m. EDT, marked a changing of the guard at CBS News Space Center. Schirra and Clarke returned to their hotel rooms; Cronkite went to sleeping quarters that had been arranged for him in one of the dressing rooms off the studio. Bill Leonard, Vice President of CBS News Programming, who had been in executive control with Salant, took Manning's place in the control room, and Wussler took a few hours break to shave, shower and change his clothes. Cross and Mary Kane, his administrative assistant, who had been in their places since 10 o'clock the previous morning, would stay in the control room throughout the early morning hours. Out on the studio floor, Joan Richman and Beth Fertik would stay at their producer and research anchor desks. They would work the 32 continuous hours.

Taking over for Cronkite as anchor man was David Schoumacher, who had manned the status desk during the first 17 hours that CBS News had been on the air. He had been given two rest periods which he spent watching Cronkite on the television set in his dressing room sleeping quarters.

Shortly after Schoumacher took over the anchor seat, he placed the second-longest long distance call since the President talked to the astronauts. He phoned the headquarters of Operation Deepfreeze in Antarctica, where he talked to Navy Commander William Hunter and Alexander Vasilev, a Russian exchange scientist. There was a great deal of static on the line, and it got worse as the conversation went on. It finally reached the point where Schoumacher could not understand a word either of the men was saying. But the things that Schoumacher remembered, and that impressed him about the talk were the Russian's sincere praise for the moon landing and the hope that future space programs could be cooperative ventures between his country and the United States.

This was a period for unwinding, much as Armstrong and Aldrin were doing in Eagle on the moon. The next two hours would be devoted to a film history of lunar exploration from the days of Aristotle, Copernicus and Galileo, through the unmanned Ranger, Lunar Orbiter and Surveyor flights to Apollos 8, 10 and 11, then the rebroadcast of some of the highlights of Armstrong's and Aldrin's moon walk.

From 6:00 to 8:00 a.m. EDT, Cross and Miss Richman drew on the resources of the film bank, using many of the pieces they had originally planned to use during the waiting period before the moon walk. Broadcast during the early morning hours when most of America was asleep, these pieces amounted to, as Schoumacher was to say later, the "best early morning programming in television history."

The film pieces, many of them months in preparation—included Correspondent Frank Kearns' profile of Dr. Louis Leakey, an archaeologist who has spent more than 40 years in Africa studying the origins of man; Correspondent Alexander Kendrick's study of "The Mystery of Stonehenge" and a history of "Great Explorations of the Past"; a Winston Burdett profile of Leonardo da Vinci's contributions to the history of flight, and a tongue-in-cheek recapitulation of the highlights of America's space program.

At 8:00 a.m. EDT, foreign reaction stories to the moon walk began to come in. First to be heard from were, as Schoumacher put it, "those Americans, who aside from our three Apollo astronauts, are farthest from home during this historic mission, our troops in Vietnam."

Correspondent Don Webster and Reporter Tony Sargent reported on the reactions of the troops in Vietnam.

Webster: *It's the middle of the night, Vietnam time. But the war keeps no regular business hours, so there were plenty of soldiers up all night, watching for Viet Cong, but listening to the progress of Apollo 11. Small television sets are now becoming commonplace in Vietnam no matter how remote. And the GIs have been following Apollo just as they* 109 *would in the States. These GIs guard a bridge on Highway 1, between Saigon and the Cambodian border. The fascination of the lunar landing is not diminished by the heat or the war. What do you think of two men landing on the surface of the moon?*

Soldier: *It really didn't impress me too much, until today. Until I was talking to a former Viet Cong who works in my company. We were trying to explain to him that the United States is putting a man on the moon. But he just refused to believe it was possible. And it really hit home at this time that the United States was accomplishing a fantastic feat.*

Sargent: *Most of the conversation here at Saigon's 3rd Field Hospital naturally revolves around such things as the war, girl friends, and going home, for these things are quite a bit closer than the moon. Sometimes of course the girl friends and the chance to go home seem almost that far away. But Apollo 11 has captured quite a bit of attention here.*

Soldier: *As a soldier, I say it means that it's another frontier that we may not have to fight in, and that would be fine. If man is smart enough, we may not have to.*

Sargent: *You have no desire yourself to go out and see what's out there?*

Soldier: *Negative.*

Mike Wallace, in London, described the excitement in Britain's capital.

Wallace: *The headline of the* Daily Mirror *here, tells it all from London. Let me read it: "The date, July 21, A.D. 1969.* MAN WALKS ON THE MOON. *Astronaut Neil Armstrong launched a new era for mankind today, when he stepped from the Lunar Module. America, land of frontiersmen, has launched a new frontier." All of the newspapers here, of course, have given banner headlines to the story. And on the television channels, all of them, this morning, they are re-playing the tapes of the astronauts' walk on the moon. But it seems everybody you talked to in London watched television all night through. And of course, there is nothing but admiration for Armstrong and Aldrin.*

110

The reaction was the same everywhere. From Daniel Schorr in Amsterdam, the word was "overwhelming." William McLaughlin in Belgrade described Apollo 11 as having taken on the elements of a "national sport." "This morning," he said, "anyone who could get near a television set did not do much work. But then, no one seems to care." And in Rome, Winston Burdett used the words "unbelievable" and "tremendous" to describe the Italians' feelings about the success of the moon walk. There was also disappointment in Rome because many of the people had gone to bed expecting to get up early to watch Armstrong and Aldrin on the moon. They awoke to find that the moon walk had taken place several hours earlier.

Meanwhile, Armstrong and Aldrin were resting on the lunar surface, and Columbia and Mike Collins were orbiting the moon some 69 miles above them. Luna 15 was keeping them company, too, circling the moon in a low orbit that had convinced most experts that the unmanned Soviet vehicle was on a reconnaissance mission, watching Apollo 11.

In the CBS News Studio at Houston, Dr. Robert Seamans, one of the pioneers of the Apollo program in the early days of NASA, and now Secretary of the Air Force, joined Bruce Morton to discuss the future of the Apollo program.

Morton: *Dr. Seamans, where should the program go from here? Obviously there are going to be more things to do on the moon.*

Seamans: *Well, I think it's important first to make use of the equipment that we have designed specifically to go to the moon. We will continue to go to the moon, and gather a maximum of scientific information. We should also use the equipment to do more work in earth orbit. The time has come when we've got to find a cheaper way of getting into orbit, so we can use space more. And this means that we can't afford to throw away all the hardware, in effect, every time we go into space. So this is one area that is going to take a lot of investigation. And depending on how that comes out, and how the other flights come out, will deter-*

111

mine the kind of space station that we will eventually want in earth orbit.

Morton: *But you think one target has got to be some sort of re-usable booster?*

Seamans: *I think it's an extremely attractive goal. I'm not convinced yet that we know exactly how to do it. So it's going to take some probing into uncharted regions, just as this present mission has gone into uncharted regions.*

Schoumacher: *Bruce, while you were taking a quick cat-nap this morning, we were discussing this problem of just how much America's space program is military versus civilian oriented. And since you have a captive down there, I thought we'd bounce this one off of Bob Seamans. Sam Phillips, former Air Force General, now Apollo Program Director, is going to be moving back into the military apparently. He's indicated to CBS News that he's resigning from the Program.*

Dr. Seamans, you were with NASA and are now back with the Air Force. Just what is the tie-up here, between the military applications of this, and the civilian uses of space?

Seamans: *First of all, NASA has received a great deal of support from the services. More than a thousand men have been assigned to the program in the course of the last 10 years. All kinds of tracking and recovery ships and so on are a very integral part of the program. And in this way the military have learned a great deal about the total operation. So that they can make use of those parts that are peculiar to military missions. Conversely, a great deal of the scientific work gives valuable background to the military in making decisions, as does the technological, administrative and management work that NASA has been engaged in.*

Morton: *Then you'd like to see this kind of spill-over between the military and the civilian agencies continue?*

Seamans: *I think it's extremely important, not only to have spill-over for reasons of national security, but spill-over into many other areas as well. I think we now have demonstrated what man can do when he really puts his mind to it and works in a cooperative way.*

112

This was the day both to contemplate the achievements of Apollo 11 and to think of the future. The subject of a manned landing on Mars came up again in a conversation Roger Mudd had with Senator Charles Percy of Illinois at the Smithsonian Institution.

Mudd: *With this success of yesterday and the apparent smoothness of operation, as you look beyond, what do you have in your mind for the United States goal in space? Do you agree with the Vice President that we should go to Mars by 1999?*

Percy: *I really feel it can be done before that. NASA is thinking in terms of the early 1980s as a real possibility. We'll be flying by to map it by 1971. We'll be getting a soft landing—instrumented by 1973. I can't imagine we'd want to wait 17 years. So that I would think with man's quest for knowledge and understanding of the universe in which we live, we'll be doing a lot to explore the eight planets and their 30 more or less moons.*

Mudd: *One of Colonel Frank Borman's grave concerns last night when he was here was keeping the team intact. Apparently that is very much on their minds. A NASA budget cut might result in the diffusion of all this expertise so to speak.*

Percy: *I think we need steadiness. These peaks and valleys, ups-and-downs, cuts and cut-backs and economy drives are very harmful to the program. You need continuity to plan ahead. We have to plan these shots years ahead.*

Mudd: *Senator, when you talk about further space exploration, do you see it as a United States-only venture, or do you envision or do you advocate some cooperation with the Russians?*

Percy: *Colonel Borman and I talked about that last night. He is a remarkable ambassador, probably one of the greatest ambassadors we've ever sent abroad. He and members of the Soviet Academy of Scientists had a talk about cooperation. And of course they're very secretive. They don't tell us what they're doing. Luna 15 showed that. But I think over a period of years we can convince the Russians of the benefit to them and to mankind for having a more open society. And to share*

113

our knowledge in space. Giving them a chance to put a man up there with us in maybe a laboratory of some sort. We need to find better avenues of opening this area of tranquility. We need them in the Middle East to help resolve that great problem. We need them to solve the very tragic war in South Vietnam. Certainly we want to start talks now on nuclear de-escalation. This is the way to find money for programs here on earth—by stopping this nuclear arms race. And if we can start to work maybe in space with the Soviet Union, possibly it will help us find a community of interest here on earth.

Lift-off from the moon's surface was just over four hours away. At 1:55 p.m. EDT, Armstrong would ignite the ascent stage engine and, using the descent stage of the LM as a launch pad, America's moon men would rocket off the lunar surface to rendezvous and dock with Columbia.

The latest report had Luna 15 still orbiting 10 miles above the lunar surface. For days there had been speculation about why the Soviet Union had chosen to launch this unmanned probe of the moon. There also was considerable interest in the Russians' reaction to Apollo 11. The Soviet Union had entered the space race well before the Americans and had taken an early but commanding lead. Sputnik I had been sent into orbit nearly four months before the United States was able successfully to orbit Explorer I, in January 1958, and cosmonaut Yuri Gagarin orbited the earth two weeks before Alan Shepard went on his suborbital flight in May 1961. The United States would wait until February 20, 1962, to place a man, astronaut John Glenn, in earth orbit. But then the United States began to catch up. Throughout the rest of Mercury and the ten-mission Gemini series, the United States showed that it knew what it was doing in space. And since Apollo 7, on October 11, 1968, everything had been going America's way. The United States had beaten Russia to the moon.

Former United Nations Ambassador Arthur Goldberg had been

in Moscow on launch day and had returned to the United States over the weekend. He discussed what Apollo 11 meant to the Russians with CBS News United Nations Correspondent Richard C. Hottelet.

Hottelet: *Justice Goldberg, what did the Apollo Mission, the launch and the enormous success of the landing stages, mean in the Soviet Union?*

Goldberg: *Well, in the Soviet Union, I think there is a great deal of interest in this launch. I cannot say it was over publicized in the initial phases. Nevertheless, to do them justice, when the launch was made it was carried in a very brief dispatch on the front pages of* Pravda *and* Izvestia *and in much greater detail, by way of analytical articles, on other pages of the newspapers. It was also announced on Soviet radio. I don't know about television. I didn't see a television, but I doubt that it was covered extensively on television. Except I understand this morning it was shown.*

Hottelet: *They had seven minutes of film.*

Goldberg: *Yes, while I was there it was covered in that way. I was entertained by many Soviet officials, both public and private. And in each instance, when I was received, entertained, talked to, the very first thing that Soviet officials did, both in the governmental community and the scientific community, was to congratulate me in behalf of the American people on this great achievement of launching a space vehicle designed to place men on the moon.*

Hottelet: *Did you get the impression they thought the United States had moved ahead of them technically?*

Goldberg: *Yes, I think they recognize that. They spoke in terms of the fact that this should not be competitive.*

Hottelet: *But in the past, when the UN sought to engage them in cooperative venture they always seemed to shy away.*

Goldberg: *They have always been very cautious about any cooperative venture. They have an obsession with secrecy. But they referred to Colonel Borman's statement several times. You will remember that Colonel Borman said as he left the Soviet Union that he looked forward* *to an early date when an American astronaut and a Soviet cosmonaut would explore space together. I remember Mr. Kuznetsov, who is First Deputy Minister, referring to Colonel Borman's statement. So I think this is something that our Government certainly should explore.*

Cronkite returned to his anchor position at 10:45 a.m. after being away for a little over seven hours. However the seven hours had not been spent sleeping. He admitted that at most he got two or three hours sleep; the rest of the time he was just too keyed up to sleep.

Shortly after 11:00 a.m., CBS News broadcast Cronkite's filmed interview with former President Johnson that had originally been scheduled for Sunday night. The interview had been filmed two weeks earlier at the Johnson ranch, where Cronkite had flown with Frank Stanton, President of CBS, Richard Salant, Gordon Manning, executive producer Burton Benjamin, producer John Sharnik and Clarence Cross. It was to be the first of a series of conversations about the Johnson years that Cronkite will have with the former President.

Cronkite: *Mr. President, in those years of '58 and '59, when you were driving this legislation through to set up a space program, did you consider that it would have political value in the campaign of '60?*

Johnson: *No, and we took precautions to see that it was not used for that purpose. We had unanimous support of members of both parties in the Congress. We had the reluctant cooperation of some of the members of the Executive Department. But when we got down to writing the legislation, and firming up the American policy, and deciding what our effort would be and how it would be administered, and the resources that would be applied to it, President Eisenhower met with us, talked to us about it, agreed to the legislation that we finally worked out with the Conference Committee, and approved the legislation. And it never got into the political arena.*

116   **Cronkite:** *In the campaign of 1960, did you have any concern that*

the top of the ticket, Senator Kennedy, wasn't as dedicated to the space program as you were? Did you see a possibility that in the new Administration you might have to sell him this program?

Johnson: *None whatever. He had supported our efforts. I never found him reluctant to agree with the recommendations we made. President Kennedy at Palm Beach in December, before we took office in January, asked me if I had any special field in which I wanted to work, and I indicated that I would be very interested in the space field. He announced to his press conference that he would ask me to head the Space Council and ask the Congress to take the President off the Space Council and put the Vice President on. That was the policy-making space group of the Government. After we did that, the President didn't lose interest in space. He met with us on a number of occasions. On April 29 I sent him a five-and-a-half-page memorandum which had unanimously been agreed upon by the Secretary of State and the Secretary of Defense, leaders in business life of this country—Mr. Brown, George Brown, Dr. Frank Stanton and Donald Cook of American Electric Power—as well as the leaders of the Senate—Senator Kerr, who was Chairman of the Space Committee, Senator Bridges and Senator Saltonstall, and other leaders, in which we all unanimously recommended certain things, and one of the paragraphs in that memorandum said that we felt that we could have a good chance—notwithstanding the Soviet exploits up to that time—of taking leadership in the space field and sending a man to the moon and bringing him back in this decade.*

In May, less than a month later—May 25, I believe—just about three weeks from the time we made our recommendations, the President finalized the message to the Congress in which he said, "Upon the advice of the Vice President and others that met with us, I believe that America should commit itself to sending a man to the moon this decade." That was an adventuresome recommendation. That took a great deal of imagination and courage, but he didn't back away from it; he faced up to it.

Cronkite: *Well, I seem to recall, Mr. President, that the unanimous decision of the Council and the recommendation to the President actually was reached by your putting it to them at a meeting, one by one around*   117 *the table, as, "Are you ready for a second-rate nation, or are you ready to endorse this program?" And that's how you got a unanimous decision.*

Johnson: *No, I don't think so. I don't think you had to beat any of these men, and we never had any problem at that time. As the effort went along in '64 and '65, some people's feet got tired and they raised the question, "Well, if we can go to the moon, why don't we take that money and do some of the things that need to be done here?" And I think that's the great significance that the space program has had. I think it was the beginning of the revolution of the '60s. I think that when we realized, after Sputnik I and II, that we had been challenged and challenged successfully, and the Soviet Union had been able to do things that up until then we had failed to do and been unable to do, and our little thirty-five-pound effort had blown up in our face. Our people are slow to start —hard to stop.*

Cronkite: *Do you think the speed with which we went for this goal of man on the moon might have been a little excessive? Would it have been better if we'd spread our $24 billion out over 20 years instead of ten, and put some of that money into other projects?*

Johnson: *I think we have enough money in this country to do all the things we need to do, and that includes space and that includes education, health and these other things. What we must have in this country is the determination to do it, and that's what we had with space.*

Reports on Luna 15 indicated that if the spacecraft kept in its present orbit, it would pass very close to Eagle at one point during its rendezvous with Columbia. The report led to increased speculation about why the Russians were keeping it in orbit for such an extended period of time.

In a way, the suspicions about the motives the Russians had in launching Luna 15 were a good example of what science fiction writer Ray Bradbury was talking about when he told Mike Wallace in London that with the moon landing we have begun "to discover we really are three billion lonely people on a small world."

Bradbury: *Jules Verne and H. G. Wells are the great-uncles and grand-fathers of all of us. When I was down in Houston a few years ago, I discovered in meeting the astronauts that they had read Jules Verne, they had read H. G. Wells, and in a few cases they had read some of my work, which made me feel very good at the time.*

Wallace: *It isn't astonishing really, because they are following a script that seems to have been written by various writers over a period of the last century.*

Bradbury: *Yes. Well, I look upon the function of the science fiction writer as being the romantic who starts things in motion, and the astronauts are the actors who come along and flesh it out and put bones inside the dream.*

Wallace: *You have said that a single invention, the rocket, is redesigning mankind. Would you elaborate on that?*

Bradbury: *Well, it's redesigning us in the following ways: I'm willing to predict tonight that by the end of the century our churches will be full again. That's redesigning mankind back in the direction of God again.*

Wallace: *Because of space travel?*

Bradbury: *Because of space travel. Because when we move out into the mystery, when we move out into the loneliness of space, when we begin to discover we really are three billion lonely people on a small world, I think it's going to draw us much closer together.*

Wallace: *Some people suggest that the very fact that we have now gone into space and have been on the other side of the moon has proved that there is no—well, the Russians have said that proves that there is no God up there.*

Bradbury: *Well, they're welcome to use any clichés they want to use. In turn, I hope to be allowed to use my clichés. I believe firmly, excitingly that we are God himself coming awake in the universe. In other words we exist on a very strange world that we know nothing about. Our theologians have tried to help us understand this. Our scientists have tried to help us understand.*

Wallace: *Can man ever feel at home elsewhere than on earth?*

Bradbury: *Yes, and he's going to make himself at home, first on the moon, then we're going off to Mars, and then we're going to build ourselves large enough ships to head for the stars, and when we do reach the nearest stars and settle there, we'll be at home in the universe. That's what the whole thing is about. This is an effort on the part of mankind to relate himself to the total universe and to live forever.*

Wallace: *Wait a minute. Live forever?*

Bradbury: *Live forever. This is an effort to become immortal. At the center of all our religions, and all of our sciences, and all of our thinking over a good period of years has been the question of death. And if we stay here on earth, we are all of us doomed, because someday the sun will either explode or go out. So in order to insure the entire race existing a million years from today, a billion years from today, we're going to take our seed out into space and we're going to plant it on other worlds. And then we won't have to ask ourselves the question of death ever again. We won't have to say, "Why existence? Why life? Why anything?" We will stop questioning in those fields.*

Shortly after noon, Cronkite broke in with the word that, according to the Jodrell Bank observatory, Luna 15 had landed on the moon. Its exact location and details as to whether it was a hard or soft landing were not available. The importance of the landing, as Cronkite said, was that it "would bring to focus if manned lunar flights are necessary."

Lift-off was one hour and 40 minutes away.

The implications of man's first landing on the moon would be discussed for years, perhaps centuries to come. Harry Reasoner had gathered four men—historian Henry Steele Commager, theologian Dr. Theodore A. Gill, psychiatrist Dr. W. Walter Menninger and author Kurt Vonnegut, Jr.—to discuss, after reflecting overnight, what they thought was most significant about the moon landing.

Commager: *I don't think it's possible to make a judgment of that kind in less than several generations, and perhaps several centuries. I suppose all Americans would assume that Columbus' exploration, and those that followed have led to something very useful to mankind. But the immediate consequences—the discovery of America—were the deaths of about 10 million Indians, and two centuries of colonial and imperial wars in Europe. And the judgment of most philosophers in the 18th Century was that the discovery of America was a great mistake. And we simply do not know, cannot know, in the nature of things, what the ultimate consequences of these will be. There's a wonderful line in Euripides, "the things men look for cometh not...and a path there is where no man thought." And that is certainly true in science as well as in history. One of the conclusions any student of science comes to, I think, quite inevitably, is that it is the by-products of the search for outer space, and of the landing on the moon, that will probably be more valuable to mankind than any immediate products that may come from the lunar landing itself.*

Reasoner: *When we got there, we found no tablets of Moses, or anything. We left a tablet signed by President Nixon. But that's not a theological subject. Are the theologians upset by this whole thing?*

Gill: *Yes, of course they'll be upset. But that's their business. What I'm hoping is that a by-product will be an atmospheric change within theology itself and the churches. I'm not greatly encouraged by the first reactions of the theologians who have been printed and who have been speaking yesterday and today. They are all continuing, it seems to me, a generally defensive stand.*

Commager: *What you want is to upset the setup.*

Gill: *Yes. So that when, for instance, the Pope speaks again, on such an occasion, he will not say how ecstatic we are about this development, but let us not forget everything else that remains to be done. So let them say how ecstatic we are, and therefore let us proceed. This has been a kind of conversion for me. I've been, as a religionist, a part of the "fine arts crowd." And it's our business to be skeptical of the scientists and the technologists. I think after yesterday I would just as soon*

121

see how many of these human personal issues involving freedom can perhaps be quantified, what a systems approach would do.

Reasoner: *That leads almost automatically to my question to Dr. Menninger. I keep reading psychiatric simplifications which say that part of our problem is that we're made to feel smaller in the age of the atom. Does this make us feel dangerously smaller and is it going to produce more schizophrenics?*

Menninger: *You know, I feel like commenting first on the religionists. This last week in Kansas there was a Pottawattami Indian gathering, and they were asking the various Indians what their impressions were about this. And there were a few who were a little concerned from their religious standpoint about this messing around with the moon.*

I think there would be two things I would respond to your comment about how small we are. I think how small we are is not just a function of the atom, but a function of communications that has dwarfed us and makes us realize how large the world is. Or how helpless we are in contrast to some of the things we can do. We can't really do anything about a tornado yet. We just have to stand helplessly by while it goes past. I think that one aspect of reaching the moon is to give us a sense of pride that we still can accomplish things. As individuals we may have been feeling more and more inadequate. As an individual, I often think "What can you or I do to stop the war in Vietnam if the President of the United States can't do it?" Or, "What can I do to stop inflation?" But obviously, collectively we can accomplish a fantastic technological achievement.

A conglomerate of individuals of private industry, of government, of academicians have all been brought together to work in remarkable harmony, after some initial upset, to accomplish this kind of grandiose achievement. And if we can do that—if we can get people to work together by finding that greater task, then there's great hope. Whether it's the young, or with deprived people, somehow we're going to be able to accomplish this. There is a discrepancy. We can reach the moon and walk on the moon, yet we can't solve the problems between people.

Reasoner: *Does it give you any hope, Mr. Vonnegut?*

Vonnegut: *Yes, I'm very excited this morning. One thing that came true last night was a prophecy of H. G. Wells. In* The Time Machine, *he predicts a time when human society will split into two distinct sorts. The basically poetic sort and engineering types. Finally the engineers become dominant. You could see this last night. Mr. Cronkite, for instance, at the time Armstrong put his foot on the moon said, "I wonder what the cynical kids are saying about this?" And this morning there was much talk about what the cynics are saying. Well, what is happening is that these are not cynics. These are different sorts of people from the engineers, and our society will split this way. Lyndon Johnson said this morning, "I believe the astronauts represent a majority of Americans." Well, a responsible President is going to have to represent the funny-looking people along with the straights.*

The time had come for Neil Armstrong and Buzz Aldrin to leave the moon. The astronauts had gone through their final checks in the LM and with Houston, making sure that every switch was in the right position. On Apollo 10, astronauts Tom Stafford and Eugene Cernan had their Abort Guidance System switch out of place, and when they fired the engine they experienced wild gyrations as the ascent stage rocketed away from the platform. It had been a lesson well learned.

The ascent stage engine would have to fire for slightly over seven minutes to raise them to the safe ten-mile limit where they could achieve orbit and could be rescued by the Command Module in the event of an emergency. This would be the first time the engine would be fired to do the job for which it had been designed, to get man off the moon. It was a tense moment, and Cronkite said to Schirra, "I don't suppose we've been this nervous about a lift-off since back in the early days of Mercury."

Banow started his animation as the countdown reached the final seconds.

Armstrong: **Forward 8, 7, 6, 5, abort stage, engine arm ascent, proceed. That was beautiful! 26, 36 feet per second up. Be advised of the pitch over. Very smooth. Very quiet ride.**

Houston: 1000 feet high, 80 feet per second vertical rise.

Cronkite: *Oh, boy! Their words "beautiful"…"very smooth"…"very quiet ride." Armstrong and Aldrin, just short of 24 hours on the moon's surface, on their way back now to rendezvous with Mike Collins orbiting the moon.*

Armstrong: A little bit of slow wobbling back and forth. Not very much thruster activity.

Schirra: *This is no problem. This is all nit-picking to describe how it flies.*

Eagle: We're going right down U.S. 1.

Capcom: Roger.

Eagle: That's Sabine off to the right now. There's Ritter out there. There it is right there. Man, that's impressive-looking, isn't it?

Armstrong and Aldrin were playing tourist, doing a little sight-seeing as they headed back into lunar orbit.

Capcom: Eagle, Houston. You're still looking mighty fine.

Eagle: About 800 to go. 700 to go. Okay, I'm opening up the main shutoffs. Ascent feed closed, pressure's holding good. Crossfeed on. 350 to go. Stand by on the engine arm. 90. Okay. Off. 50. Shutdown.

Eagle was in orbit!

The United States was one step closer to the realization of bringing man safely back to earth after landing on the moon. In less than four hours, if all went well, Armstrong and Aldrin would rendezvous and dock with Columbia.

124

An unusual thing had happened in the control room at lift-off from the moon. Banow had his filmed animation of ignition and lift-off threaded on a projector, and at the split second Armstrong fired the ascent engine Banow started the film. The LM just sat there for eight or ten seconds. The only sound in the control room, other than a few gasps, was Banow's "Oh, my God!" He immediately realized what had happened. The film should have been edited to begin the frame before ignition. Somehow, the ten seconds of lead film had been left on the roll, and when the technician froze the film on what he thought was the frame immediately before the one showing the engine fire, he actually was ten seconds away.

For just over seven minutes the animation flashed on the screen as Armstrong and Aldrin described their journey back to lunar orbit. Banow was sure that he would have to cut the simulation short to prevent having another ten seconds of engine firing after the astronauts had already shut their engines down. Then something stranger than the prolonged lift-off happened. The animation ended at the precise second Armstrong said "Shutdown." The only sound in the control room, other than a couple of "Wows," was Banow's, this time relieved, "Thank God." Two weeks later he was still trying to explain how he had started the animation ten seconds late and had it perfect, to the second, at the end.

During the minutes after lift-off Wussler had switched to Downey, California, where North American's Leo Krupp and Correspondent Terry Drinkwater simulated Mike Collins' activities in Columbia and described what he would be doing to prepare for the docking. Wussler then switched to Grumman, where test engineer Scott MacLeod and Correspondent Nelson Benton described what Armstrong and Aldrin were doing in Eagle and told why the astronauts had been unable to televise their lift-off. Although televising the lunar lift-off was scheduled for later Apollo flights, it was not made a part of the Apollo 11 mission because the camera would have drained a considerable amount of energy from the Lunar Module. No one was taking any chances.

This was a period of waiting, waiting for Eagle to perform the three individual maneuvers that would put it on an equal plane with Columbia. The first maneuver would take place at 2:53 p.m. EDT when Armstrong would fire the LM's Reaction Control System engines to place the spacecraft into an orbit some 15 to 18 miles below the Command Module.

In the period before this RCS engine firing, Cronkite checked in with the Network's affiliated stations for reactions to the moon landing and walk. In city after city—Atlanta, Dayton, Hartford, Phoenix, St. Louis, Salt Lake City, Seattle and Wichita—the story was the same: praise for Armstrong, Collins and Aldrin and joy over the accomplishments of Apollo 11.

The second crucial maneuver was performed at 3:52 p.m. EDT when the Reaction Control System engines were fired once again to place Eagle into a constant orbit 17 miles below Collins and Columbia. The firing would close the gap to 100 miles between Columbia and the trailing LM.

A new report from Sir Bernard Lovell, at the Jodrell Bank observatory with Morley Safer, indicated that Luna 15 had hit the lunar surface traveling at a high rate of speed. The prolonged silence since the landing more than five hours earlier led Sir Bernard to believe that the attempted soft landing had not been a success, that Luna 15 had crashed.

So in a two-day period there had been one notable success in moon exploration and another suspected failure. The moon, it seemed, still held some of its mystery, even after a day when its calm and silence had been violated by human footsteps and it had been exposed close hand to the world's people via television.

Harry Reasoner talked about the moon and some of the ways it might never again be the same after the events of the last 24 hours.

Reasoner: *I've been asked to say a few words about the moon at this time. The fact is, of course, I know very little about the moon. In many ways I was perfectly happy to have it that way, and it has probably been easier on me than it's going to be in the future. It's not only been easier, it's probably been more fun. The moon of my ignorance is apparently a better place than the moon of fact. It's a depressing thing about all of us that we keep reaching out for strange, unusual and interesting places, and just before we get to them, they begin to seem less strange, less unusual and less interesting. To some extent, disenchantment with the moon has already begun to set in. The moon is not as interesting to us as it once was. It appears now to be a dead rock in the void—an astronomical mistake, and not really a heavenly body at all. Just compare the facts about the moon with the lore. The truth is not nearly so strange or so interesting as the fiction.*

The moon has always had a mystical, magical quality that has made it easy for people to believe anything about it. Most moon lore differs from other astronomical fairy tales though. It usually has a sort of spooky or evil quality to it. Strange things were always happening by the light of the moon, and most of them were unpleasant. I suppose the biggest single blow that could be struck against science would be that the samples brought back from the moon turned out to be, in actual fact—green cheese. If the moon is made of green cheese, it will be encouragement to those of us whose lives are not so well ordered as the scientists' and astronauts'. It will provide some hope that we can still compete with a chance of success. Actually, I don't have much hope for the green cheese theory anymore. Our best chance now is that the moon is pure gold. If this were true, every nation on earth would drop what it's doing to go get it.

Well, anyway, I guess we're all glad we've gotten to the moon, even though we're sorry in a way that we have to burden our minds with a lot of facts, when we were perfectly happy with a lot of myths about it. I do think science has to be careful about concluding that there is not now and never has been life on the moon. Envision the possibility of visitors to this planet, a million or two million years from now. Or be specific. Set the date at a million nine hundred thousand years after we destroy ourselves with an atomic bomb, or wipe ourselves out with

127

biological warfare, or even just an epidemic of bad colds. How will the earth look then? Will there be the remains of one single initialled belt buckle to betray our past presence? It might look like the moon looks now. The man from Mars will say, "Certainly this bleak planet could never have supported life."

If everything worked on schedule, Eagle and Columbia would once again become Apollo 11 in less than an hour. Shortly after 4:30 p.m., the Reaction Control System engines would fire for the third and last time, lifting Eagle the 17 miles into orbit equal to and slightly ahead of the Command Module.

The engines were fired just before Columbia and Eagle went behind the moon.

Eagle: **We're burning.**

Columbia: **That-a-boy.**

Eagle: **Burn complete.**

Columbia: **Read, burn complete.**

Eagle: **Roger, thank you.**

The burn had been a success. When Columbia and Eagle next came around the near side of the moon at 5:23 p.m., they would have rendezvoused and would be less than 100 feet apart. Slightly over ten minutes later they would be docked.

During the waiting period, CBS News broadcast a "History of Space Journeys" narrated by Orson Welles. It included film clips from a 1902 French movie based on Jules Verne's *A Trip to the Moon*; a 1929 German movie, *Woman on the Moon*; clips from Flash Gordon and Buck Rogers films; and scenes from *Destination Moon* and *2001: A Space Odyssey*. Welles concluded his narration by saying, "Now, the moon has yielded, not merely to man's imagination, but to his actual presence. But for the science fiction writers and film makers, there remain other challenges to pose to man."

In New York Cronkite talked with Arthur C. Clarke about these new challenges.

Cronkite: *Well, Arthur, you must have been particularly interested in seeing that Orson Welles film just then, and seeing some excerpts from your own movie.*

Clarke: *It brought quite a few memories, sitting through some of those 20-, 30-, 40-year-old movies.*

Cronkite: *You know, it really is remarkable how close all of this came to reality. The only thing they didn't seem to contemplate was that the United Nations was going to be in the act, and we would have a space treaty. We wouldn't be claiming the moon. But in a sense we did, in claiming it for all mankind, as opposed to claiming it for the United States itself.*

Clarke: *I wrote a trend of that 20 years ago in my first lunar landing novel. I coined the phrase, "We shall take no frontiers into space." And I think that's the way it will be.*

Cronkite: *We've talked about where man himself is going to go, but where are you people who write science fiction going to go? Are you going to stay in space, or are you going to worry about biological problems?*

Clarke: *You know, it's a fallacy to imagine as some stupid people do, that because we've been to the moon, that's the end of science fiction. The more we discover about space, the more possibilities there are for really long-range, and yet soundly based science fiction. The early science fiction which was pure fantasy is unreadable now, but, as we discovered, more and more worthwhile. I'm looking forward to the next few years, when I absorb all this, to do my best science fiction.*

It was almost over.

At 5:23 p.m., Eagle and Columbia came wheeling around the moon, in the final phase of their docking maneuver.

Eagle: Okay, Mike. I'll get...try to get in position here, then you got it. How does the roll attitude look? I'll stop. Matter of fact, I can stop right here if you like that.

Capcom: Eagle, Houston. Middle gimbal. And you might put out to Columbia, we don't have him yet.

Eagle: They're tight. I'm not going to do a thing, Mike. I'm just letting her hold in attitude hold.

Columbia: Okay.

Cronkite: *Aldrin's job now is just to maintain position, the proper attitude so Mike can close in slowly, about a third of a mile an hour.*

Schirra: *Very, very slowly, indeed. The LM is going on automatic control now for attitude, which is pretty nearly dormant. We use the word passive, as Mike is flying his Command Module to close.*

Eagle: Okay, we're all yours. Roger.

Columbia: I'm pumping up cabin pressure.

Columbia had docked with Eagle. Then, as the astronauts talked, it became evident that the docking maneuver had not been all that they had hoped.

Columbia: That was a funny one. You know, I didn't feel it strike and then I thought things were pretty steady. I went to retract there, and that's when all hell broke loose. For you guys, did it appear to you to be that you were jerking around quite a bit during the retract cycle?

Eagle: Yes. It seemed to happen at the time I put the contact thrust to it, and apparently it wasn't centered because somehow or other I accidentally got off in attitude and then the attitude hold system started firing.

Columbia: Yes, I was sure busy there for a couple of seconds. Are you hearing me all right? I've got a horrible squeal.

Eagle: Yes, I agree with that, but we hear you okay.

Columbia: Houston, Apollo 11. Over.

Capcom: Apollo 11, Houston. Go.

It was 5:38 p.m. and Eagle and Columbia were once again Apollo 11. Less than two hours later, at 7:20 p.m., Armstrong and Aldrin would be back in the Command Module and the hatch would be closed on Eagle.

In the closing minutes Cronkite gave a "good night" and his thanks to the major domestic remotes still in operation—to Bruce Morton in Houston, Terry Drinkwater and Leo Krupp in Downey, George Herman in Flagstaff and Nelson Benton and Scott MacLeod in Bethpage.

Then he brought the historic 32 hours to an end.

Cronkite: *Man has finally visited the moon after all the ages of wishing and waiting. Two Americans with the alliterative names of Armstrong and Aldrin have spent just under a full earth day on the moon. They picked at it and sampled it, and they deployed experiments on it, and they picked away some of it to pack with them and bring home. Above the men on the moon, satellite over satellite, orbited the third member of the Apollo team, Michael Collins. His bittersweet mission was to guide and watch over the Command Service Module whose power and guidance system provided the only means of getting home, and it still does.*

Now at this point in the journey with the lunar lander reunited with the mother ship and the astronauts preparing for the rocket burn which will send them back home here, certain times and images remain that I've noted here: 4:17:40 p.m., 17 minutes and 40 seconds after four, Eastern time, yesterday — Sunday, July 20, 1969 — the moment the Lunar Module touched down on the moon's surface; 10:56 p.m. Sunday, the moment that Armstrong's foot first touched the lunar crust; and 1:54 p.m. today, the instant of lift-off from that newly named Tranquility Base camp. There were the ghostly television pictures we all saw of Aldrin and Armstrong on the moon. Armstrong's first words, "That's one small step for a man, a giant leap for mankind." And Aldrin's two-word description…"Magnificent desolation." And left behind, a plaque with the words: "HERE MEN FROM THE PLANET EARTH FIRST SET FOOT UPON THE MOON. JULY 1969 A.D. WE CAME IN PEACE FOR ALL MANKIND." And they left the flag of the United States flying there, too. Left behind, also, were hundreds of thousands of dollars' worth of cameras and hardware and equipment, discarded for the return flight; a small disc with messages microscopically reduced in size from the leaders of the world; an olive branch—symbolically at least; and two medals in memory of the three Americans and the two Russians who died in man's recent quest for the moon.

All this comes rushing back to us now as we think of the round-trip moon flight still in progress and still some critical maneuvers yet to perform. And with this flight, man has really begun to move away from the earth. But with this flight, some new challenges for mankind. A challenge to determine yet, whether in coming to the moon, we turn our century-old friend in the sky into an enemy, that we invaded, conquered, exploited and perhaps some day left as a desolate globe once more. Or will we make the most of it, as perhaps a way-station on beyond the stars. Apollo 11 still has a long way to go—and so do we.

This concludes one of the longest scheduled broadcasts—the longest in the history of television, a rather short history it is, but I think a luminous one. We've been on the air 32 hours here at CBS News Space Headquarters. I, as the man who sat here in the seat a lot of the time, sharing it with my colleague David Schoumacher, and with Wally Schirra, Arthur Clarke and our distinguished guests, our correspondents all across the nation. But more than that, the literally hundreds of technicians, engineers, associate producers, producers, writers who have produced some of the nice words that I hope we have spoken well for them here. For all of them—thanks, from Walter Cronkite, CBS

Homeward Bound
July 22, 23

The 32 hours of Sunday and Monday were behind them, but for the small group of people who gathered in Studio 41 in the first minutes of Tuesday morning, it seemed that they had never ended. They were back in the studio to cover the Transearth Injection, the rocket firing behind the moon that would send the Apollo 11 astronauts on their way home. At six o'clock, Cronkite had stayed in the studio for a piece on "The CBS Evening News," which was being anchored in his place by Harry Reasoner. Then he had gone to get something to eat, take a short nap and shower. Schirra had returned to his hotel for four or five hours sleep. Most of the other people had stayed in the building or in the area. Wussler, Mrs. Fertik and Banow, three of those involved in the post-midnight broadcast, had not dared to try to sleep. They had learned long before that one is much better off trying to stay awake after being up for a long period, when there is another broadcast coming up in six hours.

At 12:10 a.m. EDT, Apollo 11 had been given the go-ahead for Transearth Injection (TEI).

Capcom: **Apollo 11, Houston. You are go for TEI. Over.**

Collins: **Apollo 11, thank you.**

Then just before 12:30 a.m., the spacecraft disappeared around the far side of the moon. Houston had a few last words for the crew.

Capcom: **Apollo 11, Houston. One minute to LOS** [Loss of Signal]. **Go sic 'em.**

Collins: **Thank you, sir. We'll do it.**

Collins was scheduled to fire the service propulsion engine at 12:55 a.m. EDT. The burn would increase their velocity to 6188 miles per hour, the speed necessary to escape lunar orbit. The indication of success or failure would depend upon receiving a signal at precisely 1:06 a.m. The longer the wait after that time, the more the worry would grow that the engine had not fired properly and the spacecraft would be forced to make another orbit of the moon before trying the burn again.

In the minutes preceding the scheduled time of the firing, Cronkite switched to Leo Krupp and Terry Drinkwater at the North American plant. Krupp simulated what Collins would be doing at the point of ignition, and expressed every confidence in the engine's ability to perform perfectly. The maneuver had been performed in exactly the same circumstances twice before—on Apollos 8 and 10—and as Krupp said, "We have a high level of confidence in this engine."

Schirra, who had commanded the first Apollo flight, described the moment of ignition. The first of the eight burns his crew did on Apollo 7 was the one he best remembered: "Oh, you're just lying there counting down to the ignition, and suddenly there's just a big old POW! You're struggling away and suddenly you realize you're on your way. We had a lot of readouts to make almost instantaneously with the start of the engine, and I think on about the fifth burn we caught on to what was happening. The reason for this is that you don't turn a key to start this thing, or turn on an electrical generator or one of the systems that build up the current to start a spark plug. There are two fuels called hypergolics, and they burn instantaneously on contact. It's quite a slap."

The men in Mission Control had been through some exciting moments in the past two days. Another major moment was coming up. Bruce Morton, who had worked with the exception of one short nap throughout the 32 hours, described the mood there in the minutes before Acquisition Of Signal (AOS).

Morton: *This is obviously a major point in the flight for them, but the phrase that you hear from other people, the people whose shifts are finished, other NASA employees, contractors around here is, "It's really a piece of cake from here on home." That doesn't mean that they're not concerned, but it does mean that the concern here is only centered about what's new in the flight. People were worried about the landing; they were worried about problems in walking on the moon, problems in*

136

the ascent stage off the moon. Now they're back into a familiar thing, something they've done before under Apollo 8 and Apollo 10, and the feeling is, "Well, we've done that. There really isn't any good reason why it shouldn't work."

One other point I might mention. This has, I guess, been the most silent crew ever to travel in space, and in a way that's appropriate. It kind of matches the Mission Control feeling. The people here are concerned with technical points, with making sure that this or that subsystem checks out, and this really is a team effort, as people down here keep saying, and maybe it's appropriate in a way that the conversation from the moon to the ground has had so much of that character. So the people in Mission Control are now concerned, of course, but most people here think that the worst is over and things ought to go well here on in.

Schirra: *I'm willing to bet, Walter, that that back viewing room is jammed like it has been for every mission in the past. It may sound like they're a little blasé about going over something they've been over before, but they're just not showing it as much now.*

Cronkite: *Also, when you've got a critical function like this that has to work, you know there's only one chance really, it's got to go.*

Schirra: *They're on the edge of their seats just as you and I are.*

Cronkite: *First thing we get will be...*

Capcom: AOS [Acquisition Of Signal]

Cronkite: *Right on time! Right on time! They're on the way home! They have to be. We haven't heard their voice communication yet, but with the time line coming around the moon at that moment, they've got to be on the way home. It can't have speeded up enough to get around the moon at 1:06 as they did and not have the speed to get out of moon orbit and start that long ride, a quarter of a million miles.*

Voice: ...right on schedule...

Capcom: Hello, Apollo 11, Houston. How did it go? Over.

Armstrong: **Time to open up the LRL** [Lunar Receiving Laboratory] **doors, Charlie.**

Capcom: **Roger. We got you coming home. It's well stocked.**

Then Armstrong gave the statistics on the burn that had powered them out of lunar orbit.

Capcom: **Roger. Copy, Neil. Sounds good to us. All your systems look real good to us. We'll keep you posted.**

Aldrin: **Roger. Hey Charlie boy, looking good here. That was a beautiful burn. They don't come any finer.**

Armstrong, Collins and Aldrin were on their way home, and the first order of business would be a good night's sleep. Armstrong and Aldrin had had only one hour of sleep the night before on the moon, and only five the night before that. They were pretty tired.

And so were the people in Studio 41. There would be no nightcaps this night. Everyone just wanted to get to bed. The most extraordinary two days of their television lives had come to an end, and there was a feeling of great relief. But there also was an exuberance. The feeling that the job had been done right. This feeling, boosted by the earlier audience ratings for CBS News' Sunday coverage, got another lift. The figures for Monday were just as good.

The capstone to these overnight audience ratings came several weeks later when Nielsen released its nationwide audience measurements covering the flight of Apollo 11 from launch to splashdown. During the 39 hours and 25 minutes when all three television networks were covering the epic adventure, the CBS Television Network commanded a 24 percent larger average audience than NBC and a 179 percent larger average audience than ABC. And when the two astronauts walked and worked on the moon, more Americans watched them on the CBS Television Network than on the other two networks combined.

Equally gratifying to electronic journalists was that 53,520,000 television homes—94 percent of the country's total—watched some part of the three-network coverage of Apollo 11. Furthermore, television viewing had skyrocketed to substantially higher levels than normal—77 percent higher during the day and 42 percent higher at night.

But these facts would not be available for many days. This Tuesday morning the CBS News team would find satisfaction in reading the comments of television critic Jack Gould, writing in *The New York Times*:

"…there was not the slightest doubt about the superior coverage. It came from the tireless Walter Cronkite of the Columbia Broadcasting System and his colleagues. In alertness, diversity and knowhow, CBS was ahead by a wide margin.

"Mr. Cronkite for many years has been something of a one-man phenomenon in space coverage. If late yesterday afternoon his face began to show a few lines of strain, his constitution apparently is immune to fatigue. He was on the air for more hours than any other commentator, or so it seemed, and showed the greatest intimate knowledge and excitement over his assignment. He is a rare mixture, a reporter who keeps atop of developments and still reflects intensely human reactions to an unbelievable drama.

"A major CBS coup was the engagement of Walter M. Schirra, former astronaut, to sit at Cronkite's side as an expert adviser. On any number of occasions Mr. Schirra calmly explained details of space technology putting minor inconveniences in the flight into a reassuring perspective, based on his own experience.… Mr. Schirra was frequently ahead of NASA in the running narrative, and was certainly a boon in bridging the gap between the aeronautical scientist and the layman.

"Television news does appear to be undergoing subtle shifts of emphasis and tone, and CBS clearly had the best over-all grasp of the limitless implications of the moon landing. Its seminars on the scientific, religious and philosophical implications of the flight, and the reaction of militant youth, were much the best. Harry Reasoner of CBS had a diverting essay on the disappointment of the moon landing—the loss of one of mankind's most treasured sources of lore, and the discovery of forbidding dust and rock.

"…in long hours of sampling the networks, CBS…emerged with a quality of style along with an alert journalistic instinct second to none on the home screen."

Late Tuesday night, America got its first look at the astronauts since they had left the moon. Speeding toward earth at 2928 miles per hour, Apollo 11 was 180,000 miles from home when the first pictures came up at shortly after 9:00 p.m. EDT. Cronkite and Schirra were in the New York studio.

Apollo 11: Houston, Apollo 11. Over.

Capcom: Roger, go ahead, 11. Over.

Apollo 11: Are you picking up our TV signals?

Capcom: That's affirmative. We have it up on the Eidophor now. The focus is a little bit out. We see earth at the center of the screen, and still have a little white dot in the bottom of the camera apparently, and we see some land masses in the center. At least, I guess that's what it is. It's very hazy at this time on our Eidophor. Over.

Apollo 11: Believe that's where we just came from.

Cronkite: *Is that the moon or the…*

Schirra: *That's the moon. We had a label up there a moment ago saying it was the moon. Charlie Duke had a little problem with it. Our boys are doing pretty well.*

Cronkite: *I thought it was the earth at first.*

Apollo 11: It's bad enough not finding the right landing spot, but when you haven't got the right planet.

Capcom: I'll never live that down.

Apollo 11: We're making it get smaller and smaller here to make sure it really is the one we're leaving.

Capcom: That's enough, you guys.

Apollo 11: Okay, Charlie, we're getting set up for some inside pictures.

Cronkite: *Ah, that's good. That's what we want to see.*

Schirra: *See if there are really three men there?*

Then viewers were treated to one of the more interesting television shows of the Apollo program. After a few difficulties arriving at the proper light setting, a fine, clear picture came through. The color was as good as, or better than, it had ever been before.

Armstrong started the little show by giving the scientists and the world a first look at the boxes containing the lunar rocks. Then Buzz Aldrin came on with a demonstration of space food, producing a shrimp cocktail, and showing how to mix a drink and make a ham spread sandwich in a weightless environment. Then, using the ham spread tin, Buzz explained the principle of the gyroscope.

Then it was Mike Collins' turn. Speaking to "the kids at home, all kids everywhere for that matter," he demonstrated how easy it was to drink water from a spoon in zero-gravity. He then took a drink from his water gun, holding it several inches from his face as the water, in bubbles, floated toward his mouth, a little off target. He explained, "It's sort of messy. I haven't been at this very long."

Apollo 11: We'll be seeing you kids.

Capcom: Thank you from all us kids in the world, and here in Mission Control, who can't tell the earth from the moon.

Apollo 11: You want a picture now, Houston?
Capcom: That's affirmative. I refuse to bite on this one though. You tell us.

Apollo 11: Okay. This should be getting larger and if it is, it's the place we're coming up to.

Capcom: Roger.

Apollo 11: **No matter where you travel, it's always nice to get home.**

Capcom: **We concur, 11. We'll be glad to have you back.**

Then the transmission was over. It was 9:30 p.m. EDT. In just over 39 hours, Apollo 11 would splash down in the Pacific Ocean.

Wednesday, July 23

Wednesday had been a day of rest for the Apollo 11 astronauts and for the CBS News personnel still involved in the coverage of the mission. The personnel list, which had stood at approximately 350 on "Launch Day," and had swelled to nearly 1000 on "Lunar Day," had dwindled. There would be at best 225 people working on the coverage of the splashdown the next day.

With the exception of Bruce Morton in Houston and Terry Drinkwater at North American Rockwell, all the remote correspondents had returned to their regular assignments.

Cronkite, who had just finished a piece for "The CBS Evening News," was joined by Schirra at the anchor desk for the television transmission scheduled to begin shortly after 7:00 p.m. EDT. Everyone was a bit quiet and restive, reflecting on the success of the 32 hours of Sunday and Monday and thinking ahead to what they were going to do during the coverage of splashdown.

The transmission began with a picture of the Apollo 11 patch, with the American eagle swooping toward the moon with an olive branch clutched in his claws. It became evident that the astronauts were also in a reflective mood.

Armstrong: **Good evening. This is the Commander of Apollo 11. A hundred years ago, Jules Verne wrote a book about a voyage to the moon. His spaceship, Columbia [sic], took off from Florida and landed in the Pacific Ocean, after completing a trip to the moon. It seems appropriate to us to share with you some of the reflections of the crew as the modern day Columbia completes its rendezvous with the planet earth and the same Pacific Ocean tomorrow. First, Mike Collins.**

Collins: Roger. This trip of ours to the moon may have looked to you simple or easy. I'd like to say that it has not been a game. The Saturn V rocket which put us into orbit is an incredibly complicated piece of machinery. Every piece of which worked flawlessly. This computer up above my head has a 38,000-word vocabulary, each word of which has been very carefully chosen to be of the utmost value to us, the crew. This switch which I have in my hand now has over 300 counterparts in the Command Module alone. In addition to that, there are myriad circuit breakers, levers, rods, and other associated controls. The SPS engine, our large rocket engine on the aft end of our Service Module, must have performed flawlessly or we would have been stranded in lunar orbit. The parachutes up above my head must work perfectly tomorrow, or we will plummet into the ocean.

We have always had confidence that all this equipment will work, and work properly, and we continue to have confidence that it will do so for the remainder of the flight. All this is possible only through the blood, sweat and tears of a number of people. First, the American workmen who put these pieces of machinery together in the factory. Second, the painstaking work done by the various test teams during the assembly and retest after assembly. And finally, the people at the Manned Spacecraft Center, both in management, in mission planning, in flight control, and last but not least, in crew training. This operation is somewhat like the periscope of a submarine. All you see is the three of us, but beneath the surface, are thousands and thousands of others, and to all those, I would like to say thank you very much.

Aldrin: Good evening. I'd like to discuss with you a few of the more symbolic aspects of the flight of our mission, Apollo 11. But we've been discussing the events that have taken place in the past two or three days here on board our spacecraft. We've come to the conclusion that this has been far more than three men on a voyage to the moon. More, still, than the efforts of a government and industry team. More even, than the efforts of one nation.

We feel that this stands as a symbol of the insatiable curiosity of all mankind to explore the unknown. Neil's statement the other day upon first setting foot on the surface of the moon, "This is a small step for a man, but a great leap for mankind," I believe sums up these feelings very nicely. We accepted the challenge of going to the moon. The acceptance of this challenge was inevitable. The relative ease with which we carried out our mission, I believe, is a tribute to the timeliness of that acceptance.

Today, I feel we're fully capable of accepting expanded roles in the exploration of space. In retrospect, we have all been particularly pleased with the call signs that we very laboriously chose for our spacecraft, Columbia and Eagle. We've been particularly pleased with the emblem of our flight depicting the U.S. eagle bringing the universal symbol of peace from the earth, from the planet earth to the moon, that symbol being the olive branch. It was our overall crew choice to deposit a replica of this symbol on the moon. Personally, in reflecting the events of the past several days, a verse from Psalms comes to mind to me. "When I consider Thy heavens, the work of Thy fingers, the moon and the stars which Thou hast ordained, what is man that Thou art mindful of him."

Armstrong: The responsibility for this flight lies first with history and with the giants of science who have preceded this effort. Next with the American people who have through their will indicated their desire. Next, to four Administrations and their Congresses for implementing that will; and then to the agency and industry teams that built our spacecraft, the Saturn, the Columbia, the Eagle, and the little EMU, the space suit and backpack that was our small spacecraft out on the lunar surface. We would like to give a special thanks to all those Americans who built the spacecraft, who did the construction, design, the tests, and put their hearts and all their abilities into those crafts. To those people, tonight, we give a special thank you, and to all the other people that are listening and watching tonight, God bless you. Good night from Apollo 11.

The Apollo 11 crew had completed its brief transmission, one Cronkite described as "a heartwarming vote of appreciation from those three astronauts who have done the incredible, gone to the moon and walked upon it, for the people who built the spacecraft and the system which have worked so well, and the American people, as Armstrong just said, who supported the effort through the Congress of the United States."

In just over 17 hours the astronauts would be back on earth.

Splashdown
July 24

Early on the morning of July 24, the USS *Hornet*, prime recovery ship for Apollo 11, had been ordered to proceed rapidly towards Hawaii. Forced out of the intended recovery area for Apollo 11 by stormy seas, the *Hornet* moved 250 miles northwest to an area where calm seas were forecast. Nobody was taking a chance that bad weather and rough seas would conspire to cause a rough splashdown or difficult recovery for the first men who had left this planet to land and stand on another celestial body.

The scheduled time of splashdown was one hour and 21 minutes away when Cronkite and Schirra sat down at 11:30 a.m. EDT, at the anchor desk at the CBS News Space Center. If everything went as planned, the Command Module carrying astronauts Neil Armstrong, Mike Collins and Buzz Aldrin would separate from the Service Module at 12:20 p.m. EDT. Turned around, with its heat shield facing toward the re-entry corridor, the Command Module would reach a speed of 25,000 miles per hour when it passed through the earth's atmosphere. The temperature outside the spacecraft would reach 5000 degrees, and the force pressing the astronauts' bodies back against their seats would build to a pressure six times the force of gravity.

About an hour earlier, Apollo 8 astronaut Jim Lovell, one of the six men who had previously experienced the 25,000-mile-per-hour re-entry, sat in the CAPCOM's seat to reassure Mike Collins that the upcoming re-entry and splashdown were the least of his problems.

Capcom: **This is Jim, Mike. Our crew is still standing by. I just want to remind you that the most difficult part of your mission is going to come after your recovery.**

Collins: **Well, we're looking forward to all parts of it.**

Capcom: **Please don't sneeze.**

Collins: **Keep the mice healthy.**

The last part of the conversation referred to the 18 days in quarantine facing the Apollo 11 astronauts once they returned to earth. During this time the doctors would decide whether they had brought back moon germs. The mice were being kept in a sterilized chamber at the Lunar Receiving Laboratory, where they would be exposed to moon dust to see if it had any effect on them. Collins wanted to make sure that they were kept healthy.

Among the millions of people on earth waiting for Armstrong, Collins and Aldrin to return safely, three had more than a casual interest. They were the wives of the three astronauts, and CBS News Reporter Marya McLaughlin was at Mike Collins' home near Houston with a report on the women the astronauts leave behind.

Miss McLaughlin: *The astronauts' wives are almost banded together into a little tribe. The idea is to get through this mission, and then to get through those long periods when their husbands are away planning for another one. In many cases the overriding emotion is fear because these women all worry about what could happen to their husbands. Just as soon as one wife has successfully negotiated the strains and the tensions of her husband's life, she finds out that her best friend's husband will face the next one.*

Fear is not new to these women. Most of the astronauts were fliers before they became part of the astronauts' program, some on the military side, some on the civilian side, like Neil Armstrong, who tested the X-15. But the dangers of space seem endless, and there's always somebody around who says something's bound to happen because things have been going so well. And there's no getting away for these wives. NASA protects them but it surrounds them, and everyone who lives in this neighborhood works at NASA and they don't talk about anything else except space flights, the mission, and who's going to make the next one.

These women really are alone except for their children and except for each other, and they depend upon each other a great deal. Their husbands' work is dangerous and their husbands are away from home a great deal of the time, and these women run the households. They pay the bills; they raise the children; they make the major decisions. And they always dread that inevitable publicity that comes after every mission. And they worry. The astronauts' wives are always pictured as serene and calm and never concerned about a mission, but I think Joan Aldrin gave testimony the other day to the falseness of that picture when, just as Apollo 11 was blasting off Cape Kennedy, she said almost pleadingly, "Wouldn't it be wonderful if this were the countdown for the splashdown."

With splashdown an hour away, Cronkite and Schirra talked about what the minutes from re-entry to hitting the water were like. Cronkite had described them as "kind of finger-crossed moments."

Schirra: *They really are. I think when we use the word "beautiful" we use it advisedly, but I think it's emotional when you see those three parachutes out there. That is really the most beautiful sight, even though it's man-made. It's the culmination of a successful flight. It's interesting to realize I cruised back rather slowly with Apollo 7. It was only 18,000 miles an hour.*

Cronkite: *What's the scene, Wally, as you look out the window when that fireball begins, when you become a fireball?*

Schirra: *Well, the colors are quite spectacular. Of course, in the role I was in, I wasn't able to look out much. Still out of the corner of the eye I could see pastel shades of light green, light yellow, light purple, almost a violet magenta. Of course, this comes from the different elements in the heat shield, the chemicals and metals that characteristically show these colors. Copper or brass or bronze, as we normally see it, give the green. You know how the green sort of forms on a copper pipe; well that same green shows in the glow. It's quite interesting to see how the colors evolve as you pick up velocity.*

Cronkite: *Are you so busy that you can't look out?*

Schirra: *Well, we have a totally automatic re-entry capability with the computer. But we found on our first flight that it was ideal to lead the*

151

spacecraft down this corridor to a certain point; then the computer can do better. But man can do better for a part of it where you can save fuel or detect that one of the control systems may or may not be faulty, and switch to the other. The computer doesn't have that type of logic. It will just sit there and drive away and waste fuel on re-entry, and we of course must worry about that fuel. This is the fuel that's been carried in the Command Module the whole flight, never used until a check usually before separation from the Service Module. That's been dormant now for over eight days and you want to check it quite carefully.

Then came the moment for the separation of the Command Module and the Service Propulsion System Module. Astronaut Ron Evans was the CAPCOM.

Capcom: **Apollo 11, Houston. We see you getting ready for SEP. Everything looks fine down here.**

Apollo 11: **Thank you, Ron. Thank you.**

Capcom: **Apollo 11, Houston. You still look mighty fine from here. You're cleared for landing.**

Apollo 11: **We appreciate that, Ron. Roger. Gears down the lock.**

A 363-foot rocket had left the earth eight days and just under three hours earlier. It was now returning to earth as a tiny 12 by 12-foot capsule speeding towards the atmosphere.

At 12:37 p.m., Apollo 11 entered the earth's atmosphere.

Capcom: **And 11, Houston, you're going over the hill there shortly. You're looking mighty fine to us.**

Apollo 11: **See you later.**

152   Eighteen seconds after re-entry the Command Module entered a

communications blackout. For three minutes and 45 seconds, as the spacecraft dipped up and down much like a roller coaster to brake its speed, there would be no communication of any kind with the astronauts. Once through the blackout, the only direct voice communication would be from the *Hornet*.

Then the report came from Houston that the *Hornet* had made visual contact with the spacecraft. Soon after, the first voice contact was made.

Hornet: **Apollo 11, Apollo 11. This is Hornet. Hornet, over.**

Apollo 11: **Hello, Hornet, this is Apollo 11 reading you loud and clear.**

Pool correspondents Dallas Townsend of CBS News, Ron Nessen of NBC News and Keith McBee of ABC News, were on the *Hornet* to report on the splashdown and recovery and the greeting of the astronauts by President Richard M. Nixon, who had come aboard shortly after noon.

Nessen: *That fireball was the spacecraft, of course, coming down at a speed which caused the spacecraft to heat up to about 5000 degrees, and it caused the spacecraft to glow like a meteor, but actually air conditioning inside that spacecraft kept the astronauts at a comfortable 75 degrees. We should be hearing the double sonic boom as the spacecraft comes down. The astronauts at this point are dressed in their two-piece flight coveralls.*

Here's a late report saying that the Hornet's *radar room has gotten a contact on the spacecraft as it comes down.*

Cronkite: *Meanwhile we're waiting here for confirmation of drogue chute deployment, which should have come just about this second now. There we are. We have confirmation that the drogues are out.*

Nessen: *There's the double sonic boom, the sonic boom and then the echo boom off the water. That means that the drogue chutes have opened.*

Cronkite: *Obviously they're close enough—the* Hornet *has visual con-*

tact with them, can see the chutes coming down. We cannot see them with the television camera as yet.

***Houston:* Hornet reports spacecraft right on target point.**

Cronkite: *Right on target. They've still got a little bit of a ride in that slightly rough water.*

The two drogue chutes and then the main chute had opened, and Apollo 11 was floating toward the water at a point some nine miles to the port, or left side of the carrier, too far away for the television cameras to pick up the descent. The frogman team known as SWIM 1 was in the water as Neil Armstrong described his position, "1500 feet… 500 feet…100 feet." Then the word came from SWIM 1: "Splashdown. Apollo 11 has splashdown."

Within seconds, the recovery helicopters hovering overhead, the *Hornet*, Houston and Apollo 11 itself had joined in the cry of "splashdown." The television camera on the deck of the carrier picked up the bobbing capsule.

Nessen: *That was a chorus. They're back from the moon. Astronauts Armstrong, Aldrin and Collins landing in the Pacific Ocean southwest of Hawaii.*

Cronkite: *Hot dog! There they are and they're all right! Hot dog! Apollo 11 has made it!*

Townsend: *You know, Ron, it's interesting that they came down, according to my calculations, at 5:51 (12:51 p.m. EDT), which was the time planned months and months ago.*

Then the swimmers took over. Three frogmen attached flotation collars to the capsule, which had turned upside-down when it hit the water, to turn it upright. Once it was stable in the water, Mike Collins reported on the condition of the crew: "Our condition is all three excellent. We're just fine. Take your time."

Cronkite: *Now the swimmers are in the water. The swimmers will attach the flotation bags to be sure that the spacecraft is safe as it floats there, and when the hatches are open sudden waves or tilting will not flood the spacecraft. Then the swimmers will all go upwind so that when the hatch is opened they will not be contaminated if there indeed were any moon germs aboard. And one swimmer, Lt. Clancey Hatleberg, 25 years old, who now lives in San Diego but comes from Chippewa Falls, Wisconsin, will stand up, open the hatch and drop in the biological isolation garments (BIGs). Then the hatch will be closed, they'll don those garments, climb out and along with Hatleberg they will rub themselves down completely with sodium hypochlorite, which it is hoped is a moon germicide, and then be brought by helicopter back to the waiting* Hornet.

Slightly over an hour after they had splashed down in the ocean, Armstrong, Collins and Aldrin were on their way to the *Hornet* in recovery helicopter 66. As the helicopter swooped down to the carrier's deck, the ship's band played "Columbia, the Gem of the Ocean." President Nixon stood on the bridge, pointing and waving to the astronauts. The helicopter was immediately taken to an elevator and lowered to the hangar deck, where it was towed to within a few feet of the mobile quarantine facility that would be the astronauts' home until they arrived in Houston early Sunday morning. The door to the helicopter opened and Armstrong, Collins and Aldrin, looking truly like men from outer space in their dark green uniforms and masks and hoods, stepped quickly into the mobile quarantine facility (MQF), waving at the sailors on the hangar deck and the television cameras as they went.

Armstrong, Collins and Aldrin were followed by Dr. William Carpentier, who had given them a quick physical examination in the helicopter, and would now join them for the 18 days of quarantine. Joining them also was John Hirasaki, a NASA engineer who had helped design the house trailer-like facility. He would help ensure that the airtight isolation of the MQF would remain airtight until the astronauts were in Houston. One of his first duties was to get the film magazines and boxes of lunar samples out of the Command Module,   155 which had also been brought to the hangar deck. After processing the film and boxes in the MQF's decontamination lock, Hirasaki passed them to the outside to be flown to Houston. They would be in the Lunar Receiving Laboratory in Houston before the astronauts were in Hawaii.

It would be nearly an hour, following an extensive medical examination by Dr. Carpentier, before President Nixon would be able to talk to the astronauts. In the interim, the major points of the quarantine procedure were explained in profiles of Dr. Carpentier, Hirasaki and Dr. P. R. Bell, the director of the Lunar Receiving Laboratory, filmed earlier by Correspondent David Schoumacher.

Schoumacher: *It is at this point in past missions that bands started playing, the astronauts shook hands with the recovery crew and made little speeches. This time they'll be treated like they had the plague for, though the odds are extremely small, there is a chance they may.*

Besides a doctor, the only other man to come in contact with the crew up to this time will be John Hirasaki. Hirasaki had helped design the entirely self-contained transporter and will operate the power and air-conditioning systems, to be sure that total isolation is maintained.

Mr. Hirasaki, are you worried about the possibility you might wind up with some dread disease, some plague?

Hirasaki: *No, I'm not really. I think the probability of contamination is extremely remote.*

Schoumacher: *What does your wife think about this job you have?*

Hirasaki: *She thinks it's very interesting. However, she doesn't particularly agree with the extended quarantine period.*

Schoumacher: *Besides preparing the lunar samples and passing them through to the outside, Hirasaki will function as chief cook and bottle washer for the astronauts during the trip back to Houston.*

Hirasaki: *We have food preparation of airline-style meals. All we have to do is thaw the meal out, place it inside the oven, and then cook it for a pre-set time limit.*

Schoumacher: *Living inside the transporter is about like living in a* *house trailer, though the beds are a lot less comfortable. There is provision for a sixth man inside, just in case someone, a frogman for instance, should accidentally come in contact with the astronauts during the recovery operation. But other than Hirasaki, the only outsider scheduled to share the first days of quarantine with the astronauts will be a young NASA flight surgeon, Dr. William Carpentier.*

Carpentier: *We're really following them based on clinical experience. There is no new innovation, no new things that we are doing. We are following them like any physician would follow a patient for any signs of infectious disease.*

Schoumacher: *There are no particular tests that you plan to run here that are any different from the return from any mission?*

Carpentier: *Not really because all through the Apollo program we've been running the same kind of tests that we will be running after the 11 mission basically. In some of the tests we are running, we want to see if there is any change in their response to space flight, especially in the area of immunology, biochemistry, hematology. So we will be running the same protocol that we have been running on previous Apollo flights.*

Schoumacher: *Arriving at the Manned Spacecraft Center in Houston, the astronauts will spend the remainder of their 21-day isolation period in this eight-million-dollar special building, the Lunar Receiving Laboratory. Dr. P. R. Bell is in charge, not only of the astronauts, but of the samples they have brought back from the surface of the moon.*

Bell: *We get them into the vacuum laboratory and put them in some gas-free cabinets, where the cabinets are tight so that no gases leak in or out. And then these outside wrappers, which are put on to protect them, are stripped off, and all the terrestrial organisms that came from the astronauts are sterilized off the outside of these boxes. The things are guaranteed to kill terrestrial spores and microorganisms. And then these boxes are washed and dried. Then they're taken into the vacuum system and opened and the samples are taken out and looked at. We photograph them. We chip little pieces off some, most of them representative, to send down to the biological quarantine testing. And we*

157

photograph where the chips came off, and the chips themselves, and we take some small chips for physical-chemical testing to tell exactly what kind of materials and roughly what the chemical composition is, and a number of other things. But generally, most of the samples remain in the vacuum system. Then they are sealed up in vacuum-tight cans, waiting for the time when the quarantine tests are finished and we know they are safe to be distributed to other principal investigators.

For every question we answer, a thousand will grow in its place as a result of these measures. But we do hope we will get some of the basic questions answered: Whether the moon was once melted; how long these rocks sat on the surface of the moon; how long these rocks have been crystallized; if they were ever melted once or were formed as they are now; how long was it from the time they were put together as a rock? And there are a number of things of this sort we hope to find out.

After the physical examination the astronauts moved to the window to be greeted by President Nixon.

Nixon: Neil, Buzz and Mike. I want you to know that I think I'm the luckiest man in the world. I say this not only because I have the honor to be President of the United States, but particularly because I have the privilege of speaking for so many in welcoming you back to earth. I could tell you about all the messages we received in Washington. Over one hundred foreign governments, emperors, and presidents and prime ministers and kings have sent the most warm messages that we have ever received. They represent over two billion people on this earth – all of them who have had the opportunity through television to see what you have done. And then I also bring you messages from members of the Cabinet and members of the Senate and members of the House, and Space Agency.

But most important, I made a telephone call yesterday. The toll wasn't, incidentally, as great as the one I made to you fellows on the moon. I made that collect, just in case you didn't know. I called, in my view, three of the greatest ladies and most courageous ladies in the world today, your wives. And from Jan and Joan and Pat, I bring their love and their congratulations. We think that it is just wonderful that they could have participated at least through television in this return; we're only sorry they couldn't be here. And also, I've got to let you in on a little secret – I made a date with them. I invited them to dinner on the thirteenth of August, right after you come out of quarantine. It will be a State dinner held in Los Angeles. The governors of all the 50 States will be there, the ambassadors, others from around the world and in America. And they told me that you would come too. And all I want to know – will you come? We want to honor you then.

Armstrong: We'll do anything you say, Mr. President. Just anything.

Nixon: One question I think all of us would like to ask. As we saw you bouncing around in that boat out there, I wonder if that wasn't the hardest part of the journey. Did any of you get seasick?

Armstrong: No, we didn't. And it was one of the harder parts; but it was one of the most pleasant, we can assure you.

Nixon: I understand. Frank Borman says you're a little younger by reason of having gone into space. Is that right? Do you feel that way, a little younger?

Collins: We're a lot younger than Frank Borman.

Nixon: There he is, over there. Come on over Frank, so they can see you. You going to take that lying down?

Aldrin: It looks like he has aged in the last couple of days.

Nixon: Come on, Frank.

Borman: Mr. President, you know we have a poet in Mike Collins and he really gave me a hard time for using the words "fantastic" and "beautiful." And I counted them. In three minutes up there you used four "fantastics" and two "beautifuls."

Nixon: Well, just let me close off with this one thing. I was thinking, as you came down and we knew it was a success, and it had only been eight days, just a week, a long week…that this is the greatest week in the history of the world since the Creation. Because as a result of what happened this week, the world is bigger infinitely, and also as I'm going to find on this trip around the world and as Secretary Rogers will find as he covers the other countries and Asia – as a result of what you've done, the world has never been closer together   159

before. And we just thank you for that. And I only hope all of us in government, all of us in America, as a result of what you've done, can do our job a little better. We can reach for the stars just as you have reached so far for the stars.

The brief ceremony ended with a prayer by the ship's chaplain. Cronkite, picking up Frank Borman's comment that the astronauts were a little bit younger now, discussed with Schirra and Clarke the aging effect of going into space.

Cronkite: *What is this business, Arthur?*

Clarke: *The relativity effect.*

Cronkite: *The relativity effect that they're younger than they were when they left last Wednesday.*

Clarke: *The rate at which time flows is a function of your velocity. If you're traveling at very high speed, when you come back to where you started eventually you will have aged slightly less than the people you left behind. On these sorts of journeys at relatively low speeds, it's about a millionth of a second difference. But all the astronauts are in fact in the area of a millionth of a second younger than they should be. But it isn't going to be important until we start going to the stars, at speeds near that of light. Then it may add up to a difference of hundreds of years. They may come back as still young men and find that even a civilization on earth has passed away. But this sort of thing isn't to worry us until we start traveling to the stars and at speeds nearer light.*

Schirra: *There's one problem here, Walter. You do not get younger, you get less old. You don't age fast. You don't go back, unfortunately.*

Clarke: *No, you can't turn the clock back. But in principle you could come back a million years from now and be only a few years older than you were when you started.*

Cronkite: *On the other hand, whether you're younger or older, you, Wally, have seen I don't know how many more sunrises and sunsets than you are entitled to.*

160

Schirra: *About 16 per day, and let's see, after about 12 days that's quite a few of them, isn't it?*

Clarke: *Crossing the International Date Line every 90 minutes.*

Schirra: *That's the way to get younger.*

Clarke: *That's the way.*

Then Mrs. Neil Armstrong stepped to the front lawn of her Houston home to describe her feelings at the end of the mission.

Mrs. Armstrong: *I would like to say to the Presidents of the United States, President Nixon, President Johnson, President Kennedy, to all of NASA, to all the contractors that have helped make this flight successful, to the astronaut crew, to the men, the three men that made this historic flight, and to all the peoples of the world, we thank you for everything—your prayers, your thoughts, just everything. And if anyone were to ask me how I could describe this flight, I can only say that it was absolutely out of this world.*

"Man on the Moon: The Epic Journey of Apollo 11" was drawing to a close. Cronkite reflected upon what the world had seen, and how it had been changed since the morning, which now seemed so long ago, of Wednesday, July 16.

Cronkite: *Well, man's dream and a nation's pledge have now been fulfilled. The lunar age has begun. And with it mankind's march outward into that endless sky from this small planet circling an insignificant star in a minor solar system on the fringe of a seemingly infinite universe. The path ahead will be long; it's going to be arduous; it's going to be pretty doggone costly. We may hope, but we should not believe, in the excitement of today, that the next trip or the ones to follow are going to be particularly easy. But we have begun with "a small step for a man, a giant leap for mankind," in Armstrong's unforgettable words.* 161

In these eights days of the Apollo 11 mission the world was witness to not only the triumph of technology, but to the strength of man's resolve and the persistence of his imagination. Through all times the moon has endured out there, pale and distant, determining the tides and tugging at the heart, a symbol, a beacon, a goal. Now man has prevailed. He's landed on the moon, he's stabbed into its crust; he's stolen some of its soil to bring back in a tiny treasure ship to perhaps unlock some of its secrets.

The date's now indelible. It's going to be remembered as long as man survives—July 20, 1969—the day a man reached and walked on the moon. The least of us is improved by the things done by the best of us. Armstrong, Aldrin and Collins are the best of us, and they've led us further and higher than we ever imagined we were likely to go.

At 3:25 p.m. the final credits began to roll on the television screen —a visual record of the people who had contributed to making television history during the coverage of the Apollo 11 astronauts' epic adventure. It took seven minutes to complete the honor roll—the longest roster in television history.

At 3:32 p.m. EDT, Wussler and Manning shook hands around the control room, with Cross, Banow and Knox, their producer, director and associate director, then with CBS President Frank Stanton and CBS News President Richard S. Salant, both of whom had been in the control room throughout the four hours of splashdown coverage.

There was an air of celebration in the studio and editorial area. Then everyone started to drift away. The most memorable week of television in their lives had come to an end.

Voyage to the Moon

By Archibald MacLeish,
as broadcast on a wrap-up of
"Man on the Moon: The Epic Journey of Apollo 11,"
Thursday, July 24, 1969, 8:00-9:00 p.m. EDT

Presence among us,

wanderer in our skies,

dazzle of silver in our leaves and on our
waters silver,

O
silver evasion in our farthest thought—
"the visiting moon"…"the glimpses of the moon"…

and we have touched you!

From the first of time,
before the first of time, before the
first men tasted time, we thought of you.
You were a wonder to us, unattainable,
a longing past the reach of longing,
a light beyond our light, our lives—perhaps
a meaning to us…

Now
our hands have touched you in your depth of night.

Three days and three nights we journeyed,

steered by farthest stars, climbed outward,
crossed the invisible tide-rip where the floating dust
falls one way or the other in the void between,
followed that other down, encountered
cold, faced death—unfathomable emptiness…

Then, the fourth day evening, we descended,
made fast, set foot at dawn upon your beaches,
sifted between our fingers your cold sand.

We stand here in the dusk, the cold, the silence…

and here, as at the first of time, we lift our heads.
Over us, more beautiful than the moon, a
moon, a wonder to us, unattainable,
a longing past the reach of longing,
a light beyond our light, our lives—perhaps
a meaning to us…

                    O, a meaning!

over us on these silent beaches the bright
earth,
          presence among us

The Participants

EXECUTIVE PRODUCER

Robert Wussler

CO-PRODUCERS

Clarence Cross
Joan Richman

LOCATION PRODUCERS

Frank Manitzas
Jack Kelly
Sid Kaufman
Bernard Boroson
William Headline
Paul Greenberg
Peter Herford
David Fox
Jack Murphy

SEQUENCE PRODUCERS

Burton Benjamin
Ernest Leiser
Sanford Socolow
Vern Diamond
Andrew A. Rooney
Zeke Segal
Fred Warshofsky
Ronald S. Bonn
Hal Haley
Paul Soroka

DIRECTED BY

Joel Banow

INTERNATIONAL STAFF

Marshall B. Davidson
William Small
Ralph Paskman

Don Hewitt
Robert Little
Mario Biasetti
Dan Bloom
Peter Boultwood
Robert Chandler
Norman Gorin
Jeff Gralnick
Alan Harper
Arthur Kane
David Miller
Margaret Osmer
Robert Ruggiero
Joseph Tier
Robert Wilson

ASSOCIATE PRODUCERS

James O'Brien
Beth Fertik
Barry Jagoda
James Brown
Jed Duvall
Ed Freedman
Charles Gallagher
Alan Greene
Christine Huneke
Robert Jones
Arnold Labatan
Bud Lamoreaux
Patricia Lynch
Jody Porter
Joe Rothenberger
Marly Russell
Cindy Samuels
Dewey Schade
Ev Sears
Alvin Thaler
Chris Wallace
Barry Lando

LOCATION DIRECTORS

Fred Stollmack
Alvin R. Mifelow
Bob La Hendro
David Roth
Bill Barron
Dan Gingold
George Jason
Neal Finn
Joe Yaklovitch
Robert Camfiord
Robert Vitarelli
Jack Murphy
Bud Weil

ASSOCIATE DIRECTORS

Richard Knox
Arthur Bloom
Frank Bradley
James Clevenger
Joe Gorsuch
Bob Gray
Stan Green
Jim McMann
Richard Mutschler
Jan S. Rifkinson
Ken Sable
Art Spitzer
John Weaver

NEWS WRITERS

John Armstrong
John Merriman
John Mosedale

NEWS ASSOCIATES

Bob Blum
Peter Kendall

166

Leon Rice
Peter Sturtevant
Ken Witty
Bob Mead

RESEARCHERS
Mark Kramer
Margery Baker
Jim Brannon
Frances Guenette
Judy Hole
Hardy Jones
Cathy Mitchell
Howard Stringer
Rick Testa

CONSULTANTS
Richard Hoagland
Lindy Davis
Charles Friedlander
G. Harry Stine

ASSISTANTS
TO PRODUCERS
Mary Kane
Rick Boudin
Hinda Glasser
Irene Hess
Margaret Isaacs
Pamela Kossove
Susan Langley
Claire Neff
Janet Olin
Pamela Orzechowski
Carolyn Terry

DIRECTORS
OF OPERATIONS
Andrew P. Barry

Richard Sedia
Harold Sobolov

ART DIRECTOR
Hugh Gray Raisky

ENGINEERS-
IN-CHARGE
Don McGraw
Hy Badler
George Benkowsky
Wayne Brandt
Herb Gardener
Brooks Graham
Robert Heuberger
Stan Kreinik
Bob Lawson
Robert Manno
Jack O'Donnell
Walter Pile
Bob Stone
Art Tinn
John Waszak
Mal Weinges

TECHNICAL DIRECTORS
James Angerame
Guy Cornelini
Tom Delilla
Charles Donofrio
Frank Florio
Charles Franklin
Bill Guyon
Dick Hall
Bud Hvavaty
Al Kawa
George Keck
Stan Mitchel

Carl Schutzman
Harold Schutzman
Dick Shapiro
Marty Solomon
Max Streem
Stan Thorsen
Les Vaught

AUDIO
Doc Bennett
Al Bressan
A. Buckner
Herb Claudio
Norm Dewes
Tom Duffy
Jim Hargraves
Arthur Irons
Jack Katz
Sam Laine
Bud Lindquist
Fred Lopez
Mike McGrath
Romeo Quaranta
Larry Schneider
Art Shine
Jim Williams
Richard Wormsbecher

LIGHTING DIRECTORS
Stan Alper
Bob Barry
Laird Davis
William Greenfield
Ed S. Hill
Ralph Holmes
Walter Urban

SET DECORATOR
Wes Laws

167

SPECIAL EFFECTS
DIRECTORS

Bob Taylor
Henry Gordon
Mort McConnell
Neal Schatz

TITLES AND
FILM ANIMATIONS

Richard Spies
Reel III Animation

"HAL" DISPLAY

Douglas Trumbull
Trumbull Effects

PRODUCTION
SUPERVISORS

Arthur Schotz
Norman Brenner
Jim Lynch
Ray Norton
Dave Osborne
Al Rosen
Robert Sherman
Sid Sirulnick

FILM EDITORS

George Loughran
Robert H. Jegle
Irwin Dennis
Henry Neiland
Jerry Randell

STAGE MANAGERS

Rupert Baron
Bud Borgen
Don Carmichael

Willie Dahl
Chet O'Brien
Snooks O'Brien
Jim Rice
Harry Rogue
Bob Savery
Ray Sneath
Jim Wall

DIRECTOR
OF GRAPHIC ARTS

Rudi Bass

ASSOCIATE ART DIRECTOR

Ned Steinberg

GRAPHIC ARTISTS

Lowell Chereskin
Rene Gonzalez
John Huie
Joe Lagana
Billy Randell
George Smith
Billy Sunshine
Anthony Vespoli

PRODUCTION CONTROL

Michael Brigida
Steven Rader
Bernard Rozenberg
Bob Stewart
Paul Zydel

PRODUCTION LIBRARY

Dave Mlotok
Mike Chaplin
Don German

CORRESPONDENTS
NEW YORK STUDIO

Walter Cronkite
Eric Sevareid
Harry Reasoner
David Schoumacher
Charles Kuralt
Richard C. Hottelet
Alexander Kendrick

DOMESTIC LOCATIONS

Joseph Benti
Nelson Benton
Heywood Hale Broun
Terry Drinkwater
Jed Duvall
John Hart
George Herman
Marya McLaughlin
Bruce Morton
Roger Mudd
Bill Plante
Ed Rabel
Dan Rather
Bill Stout

INTERNATIONAL LOCATIONS

Mike Wallace
Winston Burdett
Marvin Kalb
Peter Kalischer
Frank Kearns
William McLaughlin
Morley Safer
Tony Sargent
Daniel Schorr
Bob Simon
George Syvertsen

168

Richard Threlkeld
Dallas Townsend
Don Webster
James Burke, BBC
Sheridan Nelson, CBC

POOL CORRESPONDENTS
Ron Nessen, NBC News
Keith McBee, ABC News
Mark Landsman, NBC News

AFFILIATE STATIONS
Ray Moore,
  WAGA-TV, Atlanta
Don Wayne,
  WHIO-TV, Dayton
Bill Haskell,
  WTIC-TV, Hartford
Bob Davies,
  KOOL-TV, Phoenix
Barry Serafin,
  KMOX-TV, St. Louis
Ted Capener,
  KSL-TV, Salt Lake City
Dick Norris,
  KSL-TV, Salt Lake City
Cliff Curke,
  KIRO-TV, Seattle
Ollie Thompson,
  KTVH-TV, Wichita

GUEST APPEARANCES BY:
Walter M. Schirra, Jr.
Arthur C. Clarke
Former President
  Lyndon B. Johnson
Vice President
  Spiro T. Agnew

Dr. Ralph Abernathy
William A. Anders
Col. Frank Borman
Ray Bradbury
Sir Francis Chichester
Dr. Alfred Chidester
Henry Steele Commager
Buster Crabbe
William Davidon
Keir Dullea
Dr. Frederick Durand
Dr. Krafft Ehricke
R. Buckminster Fuller
Rev. Theodore Gill
Dr. Thomas Gold
Arthur Goldberg
Dr. Peter Goldmark
Dr. Jon Hall
Ambassador
  Heinrich Haymerle
Robert Heinlein
Navy Commander
  William Hunter
Dr. Thor Karlstrom
Dr. Zdenek Kopal
Leo Krupp
Dr. Gerard Kuiper
Edward B. Lindaman
Jon Lindbergh
Richard Lippold
Dr. John Logsdon
Sir Bernard Lovell
Dr. Ivo Lucchitta
Scott MacLeod
Ira Magaziner
Dr. Harold Masursky
Dr. John McCauley
Dr. W. Walter Menninger
Dr. Henry Moore

Dr. Elliot Morris
Senator Charles Percy
Dr. William Pickering
Dr. David Roddy
Dr. Robert Seamans
Ambassador Samar Sen
Bob Sharp
Dr. Charles Sheldon
Don Shields
Col. Thomas Stafford
Gloria Steinem
Clyde Tambough
Newell Trask
Ambassador
  Senjin Tsuruoka
Dr. George Ulrich
Dr. Harold Urey
Alexander Vasilev
Kurt Vonnegut, Jr.
Dr. James Watson
Orson Welles
Dr. Edward Welsh
Dr. Robert Wildey
Dr. Edward Wolfe
Rev. Andrew Young

ANNOUNCER
Harry Kramer

*Editor for
10:56:20 p.m. EDT,
James Byrne*

169

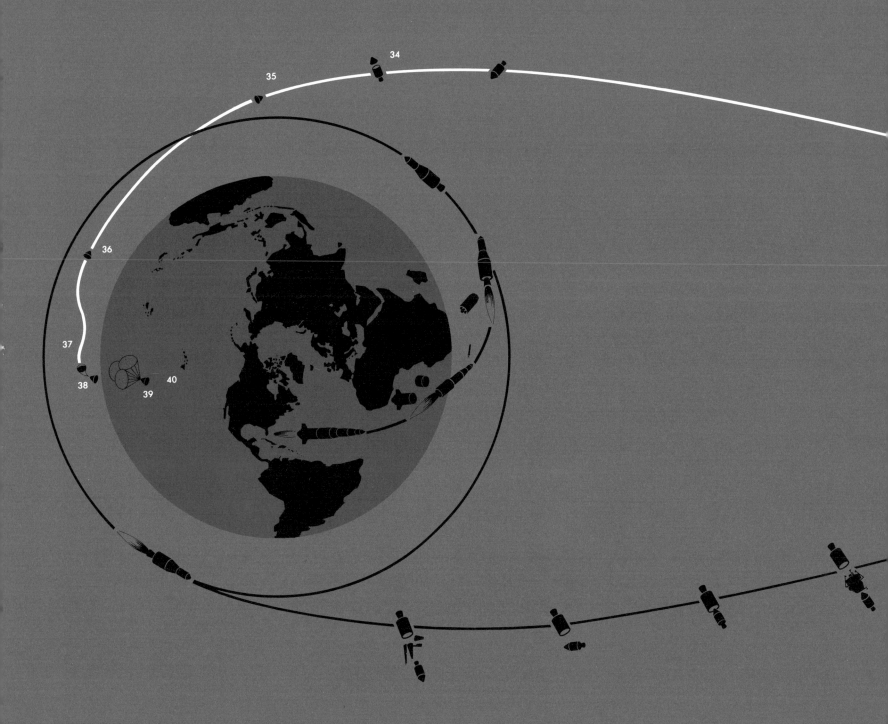